ASHEVILLE-BUNCOMBE TECHNICAL INSTITUTE

NORTH CAROLINA
STATE BOARD OF EDUCATION
DEPT. OF COMMUNITY COLLEGES
LIBRARIES

DISCARDED

NOV 20 2024

Prentice-Hall Foundation of Modern Organic Chemistry Series

KENNETH L. RINEHART, JR., Editor

Volumes published or in preparation

N. L. ALLINGER and J. ALLINGER	**STRUCTURES OF ORGANIC MOLECULES** (1965)
TRAHANOVSKY	**FUNCTIONAL GROUPS IN ORGANIC COMPOUNDS** (1971)
STEWART	**THE INVESTIGATION OF ORGANIC REACTIONS** (1966)
SAUNDERS	**IONIC ALIPHATIC REACTIONS** (1965)
GUTSCHE	**THE CHEMISTRY OF CARBONYL COMPOUNDS** (1967)
PRYOR	**INTRODUCTION TO FREE RADICAL CHEMISTRY** (1966)
STOCK	**AROMATIC SUBSTITUTION REACTIONS** (1968)
RINEHART	**OXIDATION AND REDUCTION OF ORGANIC COMPOUNDS** (1966)
DePUY and CHAPMAN	**MOLECULAR REACTIONS AND PHOTOCHEMISTRY** (1971)
IRELAND	**ORGANIC SYNTHESIS** (1969)
DYER	**APPLICATIONS OF ABSORPTION SPECTROSCOPY OF ORGANIC COMPOUNDS** (1965)
BATES and SCHAEFER	**RESEARCH TECHNIQUES IN ORGANIC CHEMISTRY** (1971)
TAYLOR	**HETEROCYCLIC COMPOUNDS**
HILL	**COMPOUNDS OF NATURE**
BARKER	**ORGANIC CHEMISTRY OF BIOLOGICAL COMPOUNDS** (1971)
STILLE	**INDUSTRIAL ORGANIC CHEMISTRY** (1968)
PAUL, RINEHART, NORMAN, and GILBERT	**X-RAY CRYSTALLOGRAPHY, MASS SPECTROMETRY, AND ELECTRON SPIN RESONANCE OF ORGANIC COMPOUNDS**
BATTISTE	**NON-BENZENOID AROMATIC COMPOUNDS**

FUNCTIONAL GROUPS IN ORGANIC COMPOUNDS

Walter S. Trahanovsky

Associate Professor of Chemistry
Iowa State University of Science and Technology

PRENTICE-HALL, INC., ENGLEWOOD CLIFFS, N.J.

To my father and mother, Nick and Ann,
Kathy, Kathy's father and mother, Walt, Jr., and Katherine Ann

© 1971 by Prentice-Hall, Inc.
Englewood Cliffs, N.J.

All rights reserved.
No part of this book may be reproduced in any form
or by any means
without permission in writing from the publisher.

Library of Congress Catalog Card Number 78-152087
Printed in the United States of America

C—13-331967-9
P—13-331959-8

PRENTICE-HALL INTERNATIONAL, INC., London
PRENTICE-HALL OF AUSTRALIA, PTY. LTD., Sydney
PRENTICE-HALL OF CANADA, LTD., Toronto
PRENTICE-HALL OF INDIA PRIVATE LIMITED, New Delhi
PRENTICE-HALL OF JAPAN, INC., Tokyo

Current Printing (last digit):
10 9 8 7 6 5 4 3 2 1

Foreword

Organic chemistry today is a rapidly changing subject whose almost frenetic activity is attested by the countless research papers appearing in established and new journals and by the proliferation of monographs and reviews on all aspects of the field. This expansion of knowledge poses pedagogical problems; it is difficult for a single organic chemist to be cognizant of developments over the whole field and probably no one or pair of chemists can honestly claim expertise or even competence in all the important areas of the subject.

Yet the same rapid expansion of knowledge—in theoretical organic chemistry, in stereochemistry, in reaction mechanisms, in complex organic structures, in the application of physical methods—provides a remarkable opportunity for the teacher of organic chemistry to present the subject as it really is, an active field of research in which new answers are currently being sought and found.

To take advantage of recent developments in organic chemistry and to provide an authoritative treatment of the subject at an undergraduate level, the *Foundations of Modern Organic Chemistry Series* has been established. The series consists of a number of short, authoritative books, each written at an elementary level but in depth by an organic chemistry teacher active in research and familiar with the subject of the volume. Most of the authors have published research papers in the fields on which they are writing. The books will present the topics according to current knowledge of the field, and individual volumes will be revised as often as necessary to take account of subsequent developments.

The basic organization of the series is according to reaction type, rather than along the more classical lines of compound class. The first ten volumes in this series constitute a core of the material covered in nearly every one-year organic chemistry course. Of these ten, the first three are a general introduction to organic chemistry and provide a background for the next six, which deal with specific types of reactions and may be covered in any order. Each of the reaction types is presented from an elementary viewpoint, but in a depth not possible in conventional textbooks. The teacher can decide how much of a volume to cover. The tenth examines the problem of organic synthesis, employing and tying together the reactions previously studied.

The remaining volumes provide for the enormous flexibility of the series. These cover topics which are important to students of organic chemistry and are sometimes treated in the first organic course, sometimes in an intermediate course. Some teachers will wish to cover a number of these books in the one-year course; others will wish to assign some of them as outside reading; a complete intermediate organic course could be based on the eight "topics" texts taken together.

The series approach to undergraduate organic chemistry offers then the considerable advantage of an authoritative treatment by teachers active in research, of frequent revision of the most active areas, of a treatment in depth of the most fundamental material, and of nearly complete flexibility in choice of topics to be covered. Individually the volumes of the Foundations of Modern Organic Chemistry provide introductions in depth to basic areas of organic chemistry; together they comprise a contemporary survey of organic chemistry at an undergraduate level.

KENNETH L. RINEHART, JR.
University of Illinois

Preface

Modern organic chemistry is a highly sophisticated, broad, and extremely important discipline. Fortunately, organic chemistry is a beautifully organized and well-developed subject, and it is thus easier to comprehend than many less extensive subjects. The goal of this book is to introduce the beginning student to the main features of organic chemistry that tie many of its facts together and lead to a systematic body of information.

The book is divided into three chapters. The first chapter presents the structure and nomenclature of organic compounds. From this section, it is hoped that the student will gain an appreciation of the basic rules for constructing formulas to represent almost all organic compounds. The important concept of functional groups, the concept which so greatly simplifies many aspects of organic chemistry, should be well understood from studying this chapter. Chapter 1 also presents the extensive and well-organized nomenclature of organic compounds. This topic is covered rather completely, and this section should be useful to the student throughout his study of organic chemistry. It should also serve as a useful reference source after the student has completed his formal training in organic chemistry.

The second chapter deals with the physical properties of organic molecules. In this chapter, there are two main objectives. First, the reader should gain a feeling for the reality of molecules. It is hoped that the student will see how certain structural features of organic molecules, especially functional groups, affect the physical properties of that substance in a reasonable and predictable fashion. The second objective of Chapter 2 is to introduce the student to the techniques that are based on physical properties of compounds, such as gas-liquid partition chromatography and the various spectroscopic methods.

In the third chapter, the student is introduced to chemical reactions of organic compounds. Again, the concept of functional groups is emphasized and the chemical interrelationships of most important functional groups are presented.

Thus, from this book, the student should gain an overall view of organic chemistry and become aware of the important features that unify many facts of organic chemistry. After finishing this book, the student should be in an excellent position to fill in details and to establish a mature understanding of many of the broad aspects of organic chemistry.

WALTER S. TRAHANOVSKY
Iowa State University of Science and Technology

The author wishes to express his gratitude to all those who influenced either in major or minor ways the writing of this book. Special thanks go to Eva Kinstle for her excellent typing of the manuscript and to Dr. C. David Gutsche for his thorough and critical reading of the manuscript.

Contents

1

STRUCTURE AND NOMENCLATURE OF ORGANIC MOLECULES   1

1.1	Introduction	1
1.2	Carbon Skeletons	4
1.3	Functional Groups that Contain Only Carbon and Hydrogen Atoms	6
1.4	Nomenclature of Hydrocarbons	8
1.5	Functional Groups that Are Attached to the Skeleton of the Molecule	21
1.6	Nomenclature for Organic Compounds that Possess Functional Groups Containing Carbon, Hydrogen, Oxygen, Nitrogen, Halogen, Sulfur, and Phosphorous Atoms	27
1.7	Nomenclature for *Cis-Trans* or Geometric Isomers	43
1.8	Nomenclature for Specification of Absolute Configuration of Asymmetric Carbon Atoms	47
	Problems	48

2

THE RELATIONSHIP BETWEEN PHYSICAL PROPERTIES AND MOLECULAR STRUCTURE 53

2.1	Introduction	53
2.2	Boiling Points	53
2.3	Melting Points	58
2.4	Solubility	64
2.5	Gas-Liquid Partition Chromatography	67
2.6	Ultraviolet-Visible (Electronic), Infrared, and Nuclear Magnetic Resonance Spectroscopy	70
	Problems	95

3

CHEMICAL INTERRELATIONS OF FUNCTIONAL GROUPS 101

3.1	Reactions of Functional Groups	101
3.2	Reactions of Alkenes (Olefins) and Alkynes (Acetylenes)—Addition Reactions	102
3.3	Preparations of Alkenes (Olefins) and Alkynes (Acetylenes)—Elimination Reactions	108
3.4	Aromatic Systems—Substitution Reactions	111
3.5	Substitution Reactions of Aliphatic Systems	114
3.6	Substitution Reactions on Unsaturated Carbon Atoms of Aliphatic Systems	117
3.7	Addition Reactions to Aldehydes and Ketones	122
3.8	Acid-Base Reactions	125
3.9	Reduction and Oxidation of Organic Compounds	129
	Problems	136

INDEX 145

1 Structure and Nomenclature of Organic Molecules

1.1 INTRODUCTION

There are two important structural features of organic molecules, a skeleton or framework and "active sites" which are part of the skeleton or are attached to it. Usually the skeleton is composed of carbon atoms and often the "active sites" are composed of atoms other than carbon atoms. An "active site" is called a *functional group* and may be defined as an atom or group of atoms of a molecule that has a special set of physical and chemical properties associated with it. The skeleton of the molecule is generally less "active" than the functional groups that it possesses, i.e., the physical and chemical properties of the whole molecule will usually be largely influenced by the functional groups that it possesses. For example, a common functional group, called a *carboxy* group, has the grouping of atoms

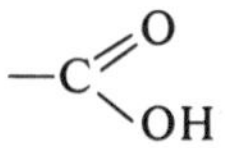

$$-C\underset{OH}{\overset{O}{\lessgtr}}$$

An example of the effect of carboxy groups on the physical properties of molecules stems from the fact that they tend to form cyclic dimers by hydrogen bonding. As a result of this favorable hydrogen-bonding situation, molecules that possess carboxy groups usually have higher melting

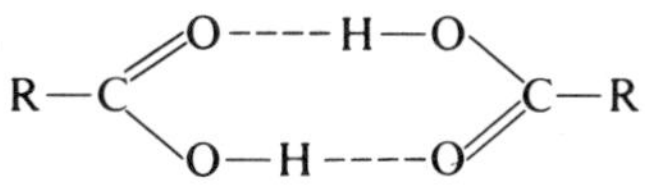

$$R-C\underset{O-H----O}{\overset{O----H-O}{\lessgtr}}C-R$$

and boiling points than they would without the carboxy group. A chemical property of most molecules that possess carboxy groups is that they

$$R-C\overset{O}{\underset{OH}{\diagdown}} + :B \rightleftharpoons \left[R-C\overset{O}{\underset{O^-}{\diagdown}} \longleftrightarrow R-C\overset{O^-}{\underset{O}{\diagdown}} \right] + HB^+$$

are acidic since the carboxy group loses a proton readily to form a resonance stabilized carboxylate anion. In fact, molecules that possess carboxy groups are called *carboxylic acids*. Thus carboxylic acids are expected and are usually found to be relatively high boiling, high melting, and acidic. Many other physical and chemical properties of carboxylic acids can be rationalized by the presence of a carboxy group.

The carboxy group is a typical functional group and from the brief discussion of some of its physical and chemical properties, it should be clear that the study of organic chemistry is greatly simplified by the concept of functional groups. In fact, organic chemistry is largely the study of the physical and chemical properties of functional groups, not individual molecules. Of course, the exact structure of the group of atoms, or *radical*, to which the functional group is attached will influence the functional group to a greater or lesser extent. This influence can be very great especially if the radical contains other functional groups.

There are three main divisions of this book. First, the structure and nomenclature of organic molecules will be discussed. From this section, the student should get a general idea of how most important organic molecules are put together and how to name these molecules. The student will be exposed to various structural features and functional groups of molecules without much attention being paid to their influence on the physical and chemical properties of molecules. This will be much like learning the appearance and names of the parts of a machine without learning the function of the parts. In the second section, the relationship between physical properties and molecular structure will be discussed. In this section as well as the section on structure of organic molecules, initial attention will be focused on the skeletons of molecules. In other words, molecules that lack functional groups or possess them within their carbon skeletons will be studied before molecules that possess functional groups attached to their carbon skeletons. This approach is necessary in the section on physical properties since it is important for the student to see how the addition of a functional group to a molecule modifies the physical properties that the molecule already possesses. Moreover, aspects of molecular structure which affect the physical properties of a molecule are often easier to understand by consideration of simpler systems, namely molecules which lack functional groups. Another reason for studying the simpler systems is that the student may more readily gain the feeling that molecules are real things and not just imagined, abstract concepts. The student should realize that many physical properties of a molecule are obviously explained by consideration of only its weight, size, shape, and flexibility.

1.1 Introduction

In the third section, the chemical interactions of functional groups will be discussed. Since the chemistry of a functional group that a molecule possesses so often dominates the chemistry of the molecule, little attention will be paid to molecules that lack functional groups. In a real sense, this third section is the heart of all of organic chemistry. Thus only a brief introduction can be presented in the space allotted. Our introduction to the chemistry of functional groups does not consist of the first part of the chemistry of organic compounds discussed in detail but is instead an overall view of the field with each topic touched on only lightly. It is hoped that the student will gain a feeling for the use of the concept of a functional group with respect to the chemistry of organic compounds, with only the most salient features of this area being learned now. Other books in the Foundations of Modern Organic Chemistry Series, such as *Ionic Aliphatic Reactions* by W. H. Saunders, Jr., *Chemistry of Carbonyl Compounds* by C. D. Gutsche, *Introduction to Free Radical Chemistry* by W. A. Pryor, *Aromatic Substitution Reactions* by L. M. Stock, *Oxidation and Reduction of Organic Compounds* by K. L. Rinehart, Jr., *Molecular Reactions and Photochemistry* by C. H. DePuy and O. L. Chapman, and *Organic Synthesis* by R. E. Ireland, should amplify these subtopics and give the student a full and detailed appreciation of the concept of functional groups in organic chemistry. In the present book, only a few mechanistic aspects of the interactions of functional groups will be mentioned, since these are introduced in R. Stewart's book in this series, *The Investigation of Organic Reactions*.

Throughout this book a main concern will be the structure of functional groups and, consequently, the structure of organic molecules. The first book in this series, *Structure of Organic Molecules* by N. L. Allinger and J. Allinger, discusses atomic and molecular structure in terms of modern wave mechanics. It is important for the student always to be aware of this meaning of structure when one discusses organic molecules. However, the aspect of structure that the organic chemist is most concerned with is the order in which atoms are attached to each other. In order to symbolize this aspect of structure, one can use simple two-dimensional formulas with lines representing a single bond (i.e., one pair of electrons). Thus, C_2H_6 and C_2H_4 can be simply represented as line drawings.

$$H-\underset{\underset{H}{|}}{\overset{\overset{H}{|}}{C}}-\underset{\underset{H}{|}}{\overset{\overset{H}{|}}{C}}-H \quad \text{and} \quad \overset{H}{\underset{H}{\diagup}}C=C\overset{H}{\underset{H}{\diagdown}}$$

On the other hand, one often wishes to show the three-dimensional aspects of structure. Since the compound containing the carbon-carbon double bond is flat, a perfectly good three-dimensional representation of it is the line drawing above. The compound containing a carbon-carbon single bond, on the other hand, must be represented by special devices such as wedges and dotted lines or circles that represent balls.

Even in the three-dimensional representation, it is unnecessary to indicate orbitals. Nevertheless, the student should always be aware of the orbital picture. He should be aware of what type of hybrid orbitals make up the various line-bonds and what their geometric relationships are. This is especially necessary when geometric isomers are considered and resonance structures are drawn, since orbitals must be properly aligned before resonance can occur.

For convenience and simplicity, most functional groups will be represented by line drawings. Nevertheless, the student should pay particular attention to the geometry of these groups and be aware of their orbital pictures.

1.2 CARBON SKELETONS

The more than two million known organic compounds and the infinite number of possible organic compounds are a result of two key facts: (1) carbon atoms can form strong bonds with other carbon atoms, and (2) carbon atoms are normally tetravalent. Knowing these two facts and using only carbon and hydrogen atoms, one can start generating structures of organic compounds. Thus, the simplest conceivable multi-carbon atom compound would be the one with two carbon atoms joined together, with each carbon atom containing three hydrogen atoms to make it tetravalent,

$$\begin{array}{c} H\ H \\ |\ | \\ H-C-C-H \\ |\ | \\ H\ H \end{array}$$

Indeed, this compound has been known for a long time and is called *ethane*. It belongs to the general class of compounds called *hydrocarbons* since it is composed of only hydrogen and carbon atoms. Ethane also belongs to the subdivision of hydrocarbons called *alkanes*. Alkanes, which are also referred to as *paraffins*, are hydrocarbons that contain only carbon-carbon single bonds. The next conceivable, and in fact well-known, hydrocarbon is *propane*, $CH_3-CH_2-CH_3$. From the formulas and from experiment the carbon skeletons of both of these molecules are seen to be linear. The tetrahedral arrangement of atoms around a carbon

1.3 Functional Groups that Contain only Carbon and Hydrogen Atoms

$$\begin{array}{c} H \\ H-C \\ | \\ CH_3 \end{array} CH_3$$

atom causes the linear chain to be bent, but nevertheless a bent line is still a line. With four carbon atoms, two conceivable carbon skeletons are possible, a linear one and a branched one. Again, both compounds are known, with the linear hydrocarbon being called *butane* and the branched hydrocarbon being called *isobutane*. Compounds such as these, which are different compounds but have the same molecular formula, are called *isomers*.

$$CH_3-CH_2-CH_2-CH_3 \qquad CH_3-\underset{\underset{CH_3}{|}}{\overset{\overset{CH_3}{|}}{C}}-H$$

Of course, one can continue this exercise (a game not unlike Tinker-Toys) and, using only carbon-carbon single bonds and forming no rings, five carbon atoms give three arrangements, six carbon atoms five arrangements, seven carbon atoms nine arrangements, ten carbon atoms 75 arrangements and 20 carbon atoms 366,319 arrangements. It is a safe bet that all the $C_{20}H_{42}$ isomers are not known! All of these arrangements involve only linear and branched chains. On paper, it is not necessary to restrict the arrangement of carbon atoms to linear and branched chains. Single rings, double rings, and other kinds of arrangements can be drawn and, as a matter of fact, nature has made most of the possibilities realities. The basic arrangements of carbon skeletons found in nature are as shown on page 6.
Of course, combinations of these basic arrangements also exist. Specific examples of these systems and rules for naming them will be presented in Section 1.4.

1.3 FUNCTIONAL GROUPS THAT CONTAIN ONLY CARBON AND HYDROGEN ATOMS

In addition to forming carbon-carbon single bonds, carbon atoms can form double and triple bonds with other carbon atoms. Hydrocarbons that contain carbon-carbon double bonds are called *alkenes* and those that contain carbon-carbon triple bonds are *alkynes*. The simplest alkene is ethene, $CH_2=CH_2$, and the simplest alkyne is ethyne, $HC\equiv CH$. Alkenes are commonly called *olefins* and alkynes are commonly called *acetylenes*. In fact, $HC\equiv CH$ is usually called "acetylene" instead of "ethyne" and is the well-known gas that is used as a fuel for torches.

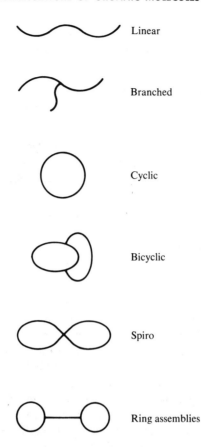

Linear

Branched

Cyclic

Bicyclic

Spiro

Ring assemblies

Carbon-carbon double and triple bonds are functional groups that are composed of only carbon and hydrogen atoms and are often found within the skeleton of the molecule. They are not always found within the major framework of the molecule since in some cases they are more reasonably considered as groups of atoms that are attached to the basic skeleton of the molecule. As functional groups they have unique sets of physical and chemical properties. This means that almost all alkenes possess certain physical and chemical properties which are characteristic of alkenes and almost all alkynes possess certain physical and chemical properties which are characteristic of alkynes. These properties will be discussed in Chapters 2 and 3.

Hydrocarbons that possess multiple bonds are said to be *unsaturated* with respect to hydrogen. Thus, unsaturated fats are those that contain carbon-carbon double or triple bonds and saturated fats contain no carbon-carbon multiple bonds.

Another very important class of hydrocarbons that are unsaturated

1.3 Functional Groups that Contain only Carbon and Hydrogen Atoms

consists of the *aromatic compounds*.* The parent aromatic compound is *benzene*, C_6H_6. Benzene is a flat molecule with the six carbon atoms arranged in a regular hexagon and a hydrogen atom attached to each carbon atom. This particular arrangement of unsaturated carbon atoms leads to

Kekulé resonance structures

an unusually stable unit and one should think of a benzene ring as a unit, not as separate carbon atoms. In fact, a benzene ring is often referred to as an *aromatic nucleus*. There are other aromatic nuclei, but the benzene ring is by far the most important one. Thus, in considering the structure of organic compounds, benzene rings can be thought of as single building blocks.

Even though all six atoms and all carbon-carbon bonds are identical, a benzene ring is often represented by a *Kekulé structure*, which depicts three double and three single carbon-carbon bonds. Justification for using Kekulé structures is that they clearly show that six π-electrons are present, which is not indicated by the circle of the six-fold symmetric symbol. When a Kekulé structure is used, it is always understood to be only one of two or more resonance forms and inequality of the carbon-carbon bonds is not implied. Resonance will be more fully discussed in Section 2.6, pp. 80-90.

Like alkenes and alkynes, aromatic compounds possess a unique set of physical and chemical properties and, thus, the aromatic nucleus could be considered to be a functional group. A saturated group of carbon atoms that might also be given the status of a functional group since it possesses characteristic physical and chemical properties is the three-carbon ring, the cyclopropane ring,

The strain energy of the small ring causes it to be much more reactive than normal saturated hydrocarbons and the small ring exhibits some unique physical properties. However, the aromatic nucleus and cyclopropane ring are not usually thought of as functional groups.

---
* N. L. Allinger and J. Allinger, *Structure of Organic Molecules*, in this series, p. 50.

8   STRUCTURE AND NOMENCLATURE OF ORGANIC MOLECULES   Chap. 1

In summary, two special arrangements of carbon atoms that are usually part of the basic skeleton of an organic compound, the carbon-carbon double and triple bonds, are considered to be functional groups. Other common and important arrangements of carbon atoms exist that possess unique sets of chemical and physical properties, but normally these are not thought of as separate functional groups.

1.4   NOMENCLATURE OF HYDROCARBONS

Exact communication is extremely important in science as in every aspect of life. Of course, exact communication is an immense problem for organic chemists, since so many organic compounds exist and the number of possible organic compounds is infinite. The organic chemist is faced with the problem of naming these compounds in such a fashion that the exact structure of the compound can be communicated from chemist to chemist. A possible system of nomenclature is one that gives every compound a unique name. This system would be comparable to the Oriental languages that essentially represent each word with a unique symbol. An alternative system of nomenclature is one that builds up the name of a given compound from smaller units in a fashion similar to western languages that represent words by a series of letters of an alphabet. The alphabet-like system of nomenclature of organic compounds should certainly be easier to learn and use in the long run. The International Union of Pure and Applied Chemistry (IUPAC) has been given the task of overseeing such systems of nomenclature. The objective of IUPAC is to bring about relatively easily used international systems of nomenclature of organic compounds that lead to names of organic compounds that convey the structures of those compounds unambiguously. To date, the IUPAC has published organic nomenclature rules regarding the nomenclature of hydrocarbons, heterocycles, and compounds that possess functional groups containing carbon, hydrogen, oxygen, nitrogen, halogen, sulfur, selenium, and/or tellurium. These rules appear in journal articles[1,2] and have also been published in book form.[3,4] They extend and revise the *Liége rules* which were published in 1933[5] by the International Union of Chemistry (IUC), which has since become the IUPAC. The Liége rules are based on the Geneva rules of 1892. A rather complete compilation of nomenclature rules is "The Naming and Indexing of Chemical Compounds," which is found in the introduction to *Chemical Abstracts*, 1962 Subject Index.[6] *Chemical Abstracts* (CA) nomenclature is essentially IUPAC nomenclature with a few modifications that are more convenient for indexing. The *Chemical Abstracts* article noted[6] includes not only IUPAC and CA nomenclature, but other nomenclature that is used commonly. The article also contains an excellent bibliography (p. 67N) which contains references to articles on nomenclature.

1.4 Nomenclature of Hydrocarbons

In this chapter, the basic approaches of the IUPAC systems are presented. The most important rules are given and illustrated by examples. Moreover, examples which involve more complex rules are presented to expose the student to the various systems used. Initially, the student will not be able to appreciate fully or comprehend these rules and examples and should thoroughly study only the basic rules and examples in each section. However, in the future the student should reread and further study each section as needed and should refer to the rules when he encounters a nomenclature problem which he cannot handle. Study of all the examples in the tables should help the student become aware of what complications can arise and will give him a clue as to how they are handled. The treatment in this chapter is not exhaustive and the reader is urged *to make use of the primary references*, especially for naming complex molecules. One should not use other textbooks for references, since their discussions are also incomplete and most of them contain errors or are misleading. Nomenclature for heterocyclic compounds will not be covered in this book but is covered in the book in this series by E. C. Taylor entitled *Heterocyclic Compounds*. Also, nomenclature rules for special classes of organic compounds, such as amino acids, will not be presented in this book.

The biggest job of the IUPAC is codifying sound practices which already exist instead of originating new nomenclature. A common misconception is that IUPAC nomenclature includes only systematic nomenclature and that other practices and systems are not sanctioned by the IUPAC. Indeed, only a few common practices are not sanctioned by the IUPAC and most compounds have two or more official IUPAC names. Even many trivial names (i.e., nonsystematic names) have been accepted as official nomenclature by the IUPAC, if they have been used extensively for many years or obviate the occurrence of complex systematic names.

There are some minor practices not used in the IUPAC or CA systems of nomenclature that are common. Usages that the organic chemist is likely to encounter in addition to those of the IUPAC or CA systems will be noted in the discussion that follows and will be referred to as "unofficial" nomenclature practices.

Nomenclature is best learned by memorization followed by use. The attentive student will soon become familiar with commonly used nomenclature practices from discussions, lectures, and literature. A workbook supplement to this series by J. G. Traynham, entitled *Organic Nomenclature: A Programmed Introduction*, offers the student a convenient means of practicing organic nomenclature.

A. Alkanes: Saturated Acyclic Hydrocarbons: In Table 1-1 are listed the names of several common saturated unbranched hydrocarbons. Also, several less common hydrocarbons are included in order to illustrate the general system. From the list, it is apparent that *-ane* is the suffix which

Table 1-1

IUPAC NAMES OF SATURATED UNBRANCHED HYDROCARBONS

Number of carbon atoms	Formula	Name	Number of carbon atoms	Formula	Name
1	CH_4	Methane	20	$CH_3(CH_2)_{18}CH_3$	Eicosane
2	CH_3CH_3	Ethane	21	$CH_3(CH_2)_{19}CH_3$	Heneicosane
3	$CH_3CH_2CH_3$	Propane	22	$CH_3(CH_2)_{20}CH_3$	Docosane
4	$CH_3(CH_2)_2CH_3$	Butane	23	$CH_3(CH_2)_{21}CH_3$	Tricosane
5	$CH_3(CH_2)_3CH_3$	Pentane	30	$CH_3(CH_2)_{28}CH_3$	Triacontane
6	$CH_3(CH_2)_4CH_3$	Hexane	50	$CH_3(CH_2)_{48}CH_3$	Pentacontane
7	$CH_3(CH_2)_5CH_3$	Heptane	100	$CH_3(CH_2)_{98}CH_3$	Hectane
8	$CH_3(CH_2)_6CH_3$	Octane	154	$CH_3(CH_2)_{152}CH_3$	Tetrapentaconta-hectane
9	$CH_3(CH_2)_7CH_3$	Nonane			
10	$CH_3(CH_2)_8CH_3$	Decane			
11	$CH_3(CH_2)_9CH_3$	Undecane			
12	$CH_3(CH_2)_{10}CH_3$	Dodecane			
13	$CH_3(CH_2)_{11}CH_3$	Tridecane			
14	$CH_3(CH_2)_{12}CH_3$	Tetradecane			
15	$CH_3(CH_2)_{13}CH_3$	Pentadecane			

indicates an alkane. The trivial names of the first four alkanes have been retained. The higher alkanes are composed of names derived from the Greek or Latin numerals followed by -ane. Note that *undecane* is used instead of *hendecane*. At least the names of the first twelve alkanes should be memorized since they are frequently encountered in the literature of organic chemistry.

Saturated unbranched hydrocarbon radicals that are derived from unbranched alkanes by removal of a hydrogen from a terminal carbon atom are named by replacing the suffix -*ane* by -*yl*. These radicals are called *alkyl* radicals. A singly-branched alkane is named by prefixing the name of the substituent alkyl group to that of the longest chain present. The position of the alkyl group is indicated by a number; the longest chain or parent is numbered from one end to the other in the direction that leads to the lowest number for the alkyl group. A multiply-branched alkane is named in the same way. The direction of the numbering of the parent is chosen by comparison of the two series of numbers of the substituents term by term. The direction that gives rise to the series that has the lower number at the first point of difference is used. Study of the names of the first three branched alkanes shown in Table 1-2 will make these rules easy to understand and apply.

In addition to systematic nomenclature for branched alkanes, trivial or semi-trivial names have been retained by the IUPAC for the following

1.4 Nomenclature of Hydrocarbons

Table 1-2

IUPAC NAMES OF VARIOUS BRANCHED ALKANES

1. $\overset{1}{C}H_3\overset{2}{C}H_2\overset{3}{C}H\overset{4}{C}H_2\overset{5}{C}H_2\overset{6}{C}H_3$ 3-Methylhexane
   $\qquad\;\;|$
   $\qquad CH_3$

2. $\overset{6}{C}H_3\overset{5}{C}H_2\overset{4}{C}H\overset{3}{C}H_2\overset{2}{C}H\overset{1}{C}H_3$ 2,4-Dimethylhexane
   $\qquad\;\;|\qquad\;\;|$
   $\qquad CH_3\;\;\;CH_3$

3. $\overset{10}{C}H_3\overset{9}{C}HCH_2\overset{7}{C}H\overset{6}{C}H_2\overset{5}{C}H_2\overset{4}{C}H_2\overset{3}{C}H_2\overset{2}{C}(CH_3)\overset{1}{C}H_3$ 2,2,7,9-Tetramethyldecane
   with CH_3 branches at 9, 7, 2, 2

4. $\overset{9}{C}H_3\overset{8}{C}H_2\overset{7}{C}H_2\overset{6}{C}H_2\overset{5}{C}H\overset{4}{C}H_2\overset{3}{C}H_2\overset{2}{C}H_2\overset{1}{C}H_3$ 5-Isobutylnonane
   $\qquad\qquad\qquad\;\;|$
   $\qquad\qquad\quad\;\;CH_2$
   $\qquad\qquad\qquad\;\;|$
   $\qquad\qquad\quad CH(CH_3)_2$

5. $\overset{1}{C}H_3\overset{2}{C}H_2\overset{3}{C}H_2\overset{4}{C}H\overset{5}{C}H_2\overset{6}{C}H_2\overset{7}{C}H_3$ 4-*tert*-Butylheptane
   $\qquad\qquad\;\;|$
   $\qquad\quad\;\;C(CH_3)_3$

6. $\overset{8}{C}H_3\overset{7}{C}H_2\overset{6}{C}H_2\overset{5}{C}H-\overset{4}{C}H\overset{3}{C}H_2\overset{2}{C}H\overset{1}{C}H_3$ 2-Methyl-5-ethyl-4-propyloctane
   $\qquad\;\;|\quad\;\;|\quad\;\;|$ (increasing complexity)
   $\quad\;\;CH_2\;CH_2\;CH_3$ 5-Ethyl-2-methyl-4-propyloctane
   $\qquad\;\;|\quad\;\;|$ (alphabetical)
   $\quad\;\;CH_3\;CH_2$
   $\qquad\qquad\;|$
   $\quad\qquad\;CH_3$

(*Note:* 4-Ethyl-5-isobutyloctane is not used since the chain having the greatest number of side chains is chosen as part of the parent chain if there is a choice between two chains of equal length.)

7. $\overset{11}{C}H_3\overset{10}{C}H_2\overset{9}{C}H_2\overset{8}{C}H_2\overset{7}{C}H_2\overset{6}{C}H\overset{5}{C}H_2\overset{4}{C}H\overset{3}{C}H_2\overset{2}{C}H_2\overset{1}{C}H_3$
   with side chain $\overset{1}{C}H_3-\overset{2}{C}H\overset{3}{C}H_2\overset{4}{C}H_3$ at position 6
   $\qquad\qquad\;\;|$
   $\qquad\qquad CH_3$
   and CH_2CH_3 at position 4

   6-(1,3-Dimethylbutyl)-4-ethylundecane

8. $\overset{1}{C}H_3\overset{2}{C}H_2\overset{3}{C}H_2\overset{4}{C}H_2\overset{5}{C}H_2\overset{6}{C}H\overset{7}{C}H_2\overset{8}{C}H_2\overset{9}{C}H_2\overset{10}{C}H_2\overset{11}{C}H_2\overset{12}{C}H_2\overset{13}{C}H\overset{14}{C}H_2\overset{15}{C}H_2\overset{16}{C}H_2\overset{17}{C}H_2\overset{18}{C}H_2\overset{19}{C}H_2\overset{20}{C}H_3$
   with $CH_3CHCH_2CH_2CH_2CH_3$ at position 13
   and $CH_2CH_2CH_2CHCH_3$ with CH_3 at position 6

   13-(1-Methylpentyl)-6-(4-methylpentyl)eicosane

four unsubstituted hydrocarbons:*

$(CH_3)_2CHCH_3$	isobutane
$(CH_3)_2CHCH_2CH_3$	isopentane
$(CH_3)_4C$	neopentane
$(CH_3)_2CHCH_2CH_2CH_3$	isohexane

These names have been derived from the following prefixes which are used in unofficial nomenclature:

Prefix	Meaning of prefix
normal-	Hydrocarbon is unbranched
iso-	Hydrocarbon contains $(CH_3)_2CH-$ and no other branches
neo-	Hydrocarbon contains $(CH_3)_3C-$ and no other branches

Trivial or semi-trivial names have also been accepted by the IUPAC for the following unsubstituted radicals:

$(CH_3)_2CH-$	Isopropyl
$(CH_3)_2CHCH_2-$	Isobutyl
$CH_3CH_2\underset{\underset{CH_3}{\|}}{C}H-$	sec-Butyl
$(CH_3)_3C-$	tert-Butyl
$(CH_3)_2CHCH_2CH_2-$	Isopentyl†
$(CH_3)_3CCH_2-$	Neopentyl†
$CH_3CH_2\underset{\underset{CH_3}{\|}}{\overset{\overset{CH_3}{\|}}{C}}-$	tert-Pentyl†
$(CH_3)_2CHCH_2CH_2CH_2-$	Isohexyl

The prefixes sec- and tert- are abbreviations for "secondary" and "tertiary." Using unofficial nomenclature, chemists often abbreviate these terms by s- and t-.

Other univalent branched radicals are named by prefixing the name of the side chain to that of the unbranched radical. The position of the

---

*Chemical Abstracts refers to these four compounds as 2-methylpropane, 2-methylbutane, etc.

†The name *amyl* is often used to replace *pentyl* in unofficial nomenclature, especially in the case of "isoamyl."

1.4 Nomenclature of Hydrocarbons

side chain is indicated by numbering the carbon atom with the free valence as one.

If two or more different kinds of side chains are present, the names of the side chains are placed either (1) in order of increasing complexity or (2) in alphabetical order. The radical containing the greater number of carbon atoms is more complex. Radicals are alphabetized by their first letter even if complex. Further rules for ordering radicals can be found in the IUPAC article.[1,3] The examples in Table 1-2 illustrate these rules.

B. Alkenes and Alkynes: Unsaturated Acyclic Hydrocarbons: *Alkenes* are hydrocarbons that contain one carbon-carbon double bond. Hydrocarbons that contain more than one double bond are *alkadienes, alkatrienes,* etc. Hydrocarbons that contain one or more carbon-carbon triple bonds are called *alkynes, alkadiynes,* etc. Unbranched unsaturated hydrocarbons are named by replacing the *-ane* of the name of the corresponding alkane with *-ene, -yne*, etc., and signifying the position of the unsaturation by numbers in front of the name. If there is only one double or triple bond, the chain is so numbered that the unsaturation receives the lower number. In cases where several points of unsaturation exist, the two series of numbers are compared and the series that has the lower number at the first point of difference is used. Hydrocarbons that contain both a double bond and a triple bond are given the ending *-enyne*. The chain is numbered so that the lowest numbers result. If either the double or triple bond can receive the lowest number, the double bond is given the lowest number.

Alkenes are commonly called *olefins* and *alkynes* are commonly called *acetylenes*.

Trivial names that the IUPAC has accepted are:

$CH_2{=}CH_2$	Ethylene
$CH_2{=}C{=}CH_2$	Allene
$HC{\equiv}CH$	Acetylene

Branched unsaturated hydrocarbons are named as derivatives of the longest unbranched carbon chain that contains the maximum number of multiple bonds. The parent name is obtained from this carbon chain. Univalent radicals derived from unsaturated hydrocarbons are given the endings *-enyl, -ynyl*, etc. Divalent radicals generated by removal of two hydrogen atoms from one carbon atom are given the ending *-ylidene*.

The IUPAC has also retained the names:

$CH_2{=}CH{-}$	Vinyl
$CH_2{=}CH{-}CH_2{-}$	Allyl
$CH_2{=}\underset{\underset{CH_3}{\vert}}{C}{-}$	Isopropenyl

In Table 1-3 are given several examples of names of unsaturated hydrocarbons that illustrate these rules.

Table 1-3
IUPAC NAMES OF VARIOUS ALKENES AND ALKYNES

1.	CH₃CH=CHCH₃	2-Butene
2.	CH₃CH=CHCH₂C≡CCH₃	2-Hepten-5-yne
3.	CH₃CH=CHCH₂C≡CH	4-Hexene-1-yne

4.	CH₃—CH₂—CH—CH=CH₂
	        |
	        CH₂—CH₂—CH₂—CH₃
		3-Ethyl-1-heptene

5.	CH₃—CH=C—C≡C—CH₂CH₃
	        |
	        CH₂CH₂CH₂CH₂CH₃
		3-Pentyl-2-hepten-4-yne

6.	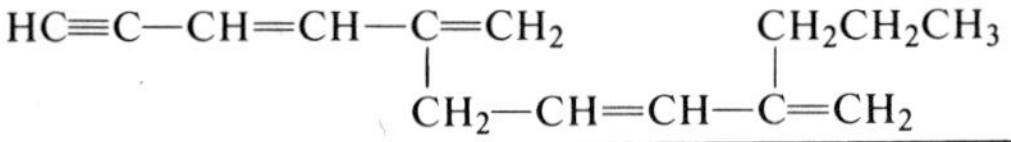 HC≡C—CH=CH—C=CH₂ CH₂CH₂CH₃
	                          | |
	                          CH₂—CH=CH—C=CH₂
		6-Methylidene-2-propyl-1,3,7-decatrien-9-yne

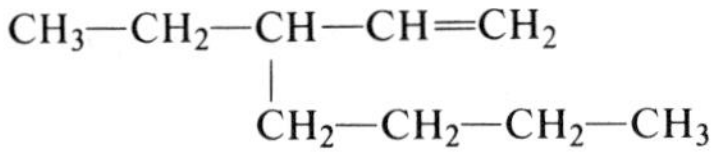

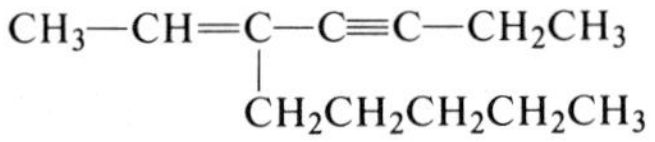

C. Cyclic Hydrocarbons: Cyclic hydrocarbons with no branches are named by adding the prefix *cyclo-* to the name of the open chain hydrocarbon that contains the same number of carbon atoms. Frequently, only line drawings are used to represent rings. Univalent radicals derived from cyclic hydrocarbons are named in the obvious way with the carbon atom with the free valence being numbered 1. The names in Table 1-4 illustrate these rules.

Table 1-4
IUPAC NAMES OF NONAROMATIC CYCLIC HYDROCARBONS

1.	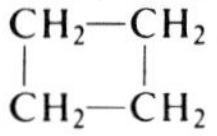 CH₂—CH₂ | | CH₂—CH₂	Cyclobutane
2.	 (pentagon)	Cyclopentene
3.	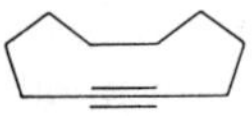 (decagon with triple bond)	Cyclodecyne
4.	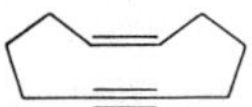 (decagon with double bond)	1-Cyclodecen-6-yne
5.	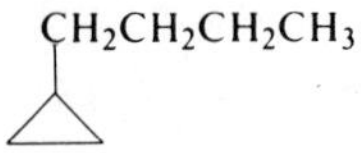 CH₂CH₂CH₂CH₃ on cyclopropane	1-Cyclopropylbutane
6.	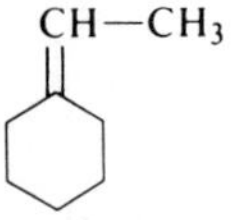 CH—CH₃ on cyclohexane	Ethylidenecyclohexane
7.	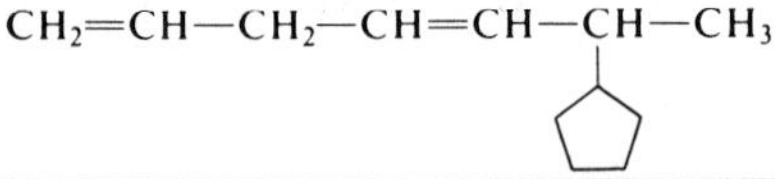 CH₂=CH—CH₂—CH=CH—CH—CH₃ with cyclopentyl	6-Cyclopentyl-1,4-heptadiene

1.4 Nomenclature of Hydrocarbons

Many trivial names have been accepted as official nomenclature for aromatic compounds. The most common compounds are shown in Table 1-5. Notice that the very common phenyl radical is symbolized by Ph-, φ-, or C_6H_5- as well as by a regular hexagon with three lines or a circle inside which represents the π-electrons.

Table 1-5
IUPAC NAMES OF COMMON AROMATIC COMPOUNDS AND RADICALS

Structure	Name
(hexagon with three lines), (hexagon with circle)	Benzene
(phenyl structures), Ph-, φ-, C_6H_5-	Phenyl
ortho-disubstituted ring	ortho-Phenylene
Ph-CH(CH$_3$)$_2$	Cumene
Ph-$\overset{\alpha}{C}H=\overset{\beta}{C}H_2$, φ-CH=CH$_2$	Styrene
Ph-CH$_3$	Toluene
meta-(CH$_3$)$_2$ ring	meta-Xylene (m-xylene)
1,3,5-trimethyl ring (CH$_3$, CH$_3$, CH$_3$)	Mesitylene
Ph-CH$_2$—	Benzyl
φ-CH=CH—CH$_2$—	Cinnamyl
Ph-$\overset{\beta}{C}H_2\overset{\alpha}{C}H_2$—	β-Phenethyl
φ_3C—	Trityl or triphenyl- methyl

The position of substituents on benzene rings is indicated by numbers except that *ortho-* (*o-*), *meta-* (*m-*), and *para-* (*p-*) may be used for 1,2-, 1,3-, and 1,4-disubstituted derivatives, though not for multi-substituted derivatives.

D. Bridged Hydrocarbons: Many hydrocarbon systems are composed of two rings which have two or more atoms in common. These systems are given the name of the straight chain hydrocarbon which contains the same number of carbon atoms preceded by the prefix *bicyclo-*. Between the prefix *bicyclo-* and the parent name are inserted three numbers which indicate the number of atoms in each bridge. These numbers are placed in decreasing order, enclosed in brackets, and separated by periods. These hydrocarbon systems are numbered by starting with a bridgehead and numbering around the longest bridge to the other bridgehead, continuing back to the first bridgehead along the second longest bridge, and finishing along the shortest bridge. For unsaturated bicyclic systems, the unsaturation is given the lowest number only if there is a choice. Also, radicals retain the numbering of the hydrocarbon and the point or points of attachment are given the lowest possible numbers if there is a choice. The examples in Table 1-6 illustrate these and other rules. Frequently, the structures of bicyclic systems are drawn in three dimensions.

Table 1-6

IUPAC NAMES OF BICYCLIC HYDROCARBONS

	Formula	Three-dimensional line drawing	Name
1.	$^6CH_2-\overset{H}{\underset{|}{\overset{|}{C}}}-^2CH_2$ $\vert$ $^7CH_2$ $^5CH_2-\underset{4}{\overset{|}{C}}-\underset{3}{CH_2}$ H		Bicyclo[2.2.1]heptane (norbornane)[a]
2.	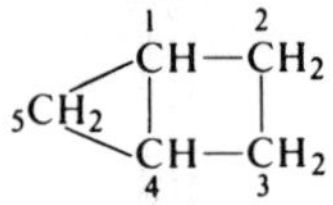 $_5C\overset{1}{\underset{}{H_2}}\overset{CH-^2CH_2}{\underset{CH-CH_2}{\diagdown}}$ $\underset{4}{}\underset{3}{}$	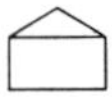	Bicyclo[2.1.0]pentane
3.	$^6CH_2\overset{^1CH}{\underset{|}{\diagup}}\overset{^2CH_2}{\underset{|}{\diagdown}}$ $^7CH_2$ $^8CH_2$ $_5CH_2\underset{\underset{4}{CH}}{\diagdown}\overset{}{\diagup}\underset{3}{CH_2}$		Bicyclo[2.2.2]octane
4.	$\overset{7}{CH}-\overset{1}{CH}-\overset{2}{CH_2}$ $\parallel\overset{8}{\underset{|}{CH_2}}\overset{3}{\underset{|}{CH_2}}$ $\underset{6}{CH}-\underset{5}{CH}-\underset{4}{CH_2}$		Bicyclo[3.2.1]oct-6-ene
5.	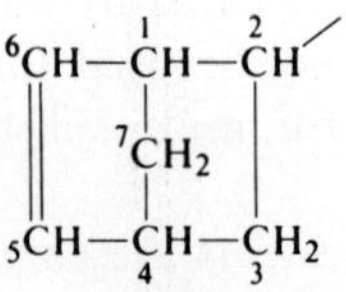 $^6CH-\overset{1}{CH}-\overset{2}{CH}$ $\parallel^7CH_2$ $_5CH-\underset{4}{CH}-\underset{3}{CH_2}$	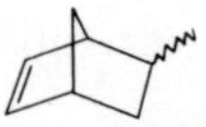	Bicyclo[2.2.1]hept-5-en-2-yl (5-norbornen-2-yl)[a]
6.	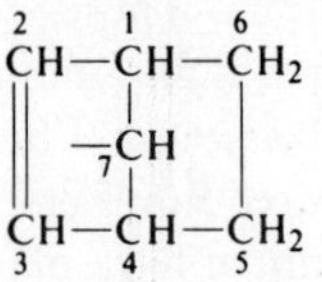 $\overset{2}{CH}-\overset{1}{CH}-\overset{6}{CH_2}$ $\parallel_7CH$ $\underset{3}{CH}-\underset{4}{CH}-\underset{5}{CH_2}$	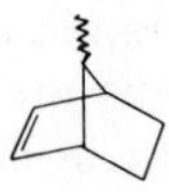	Bicyclo[2.2.1]hept-2-en-7-yl (2-Norbornen-7-yl)[a]
7.	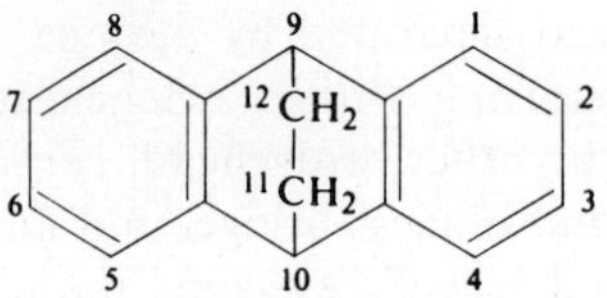 (anthracene numbering with $^{12}CH_2$ and $^{11}CH_2$ bridge)	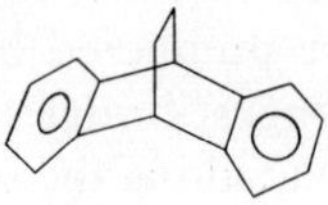	9,10-Dihydro-9,10-ethanoanthracene

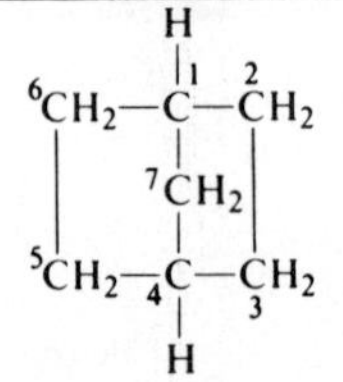

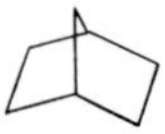

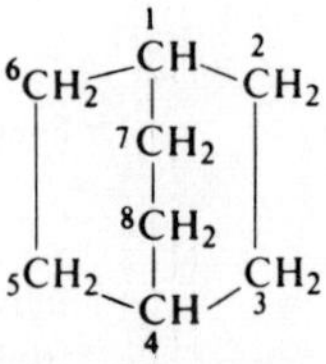

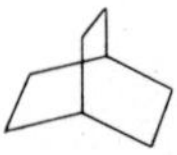

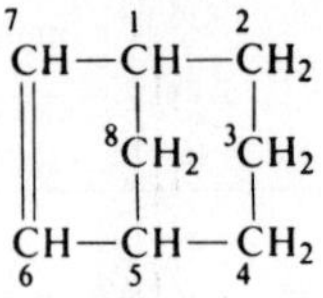

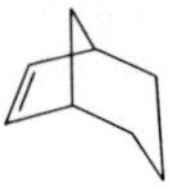

1.4 Nomenclature of Hydrocarbons

Table 1-6 (*continued*)

IUPAC NAMES OF BICYCLIC HYDROCARBONS

	Formula	Three-dimensional line drawing	Name
8.	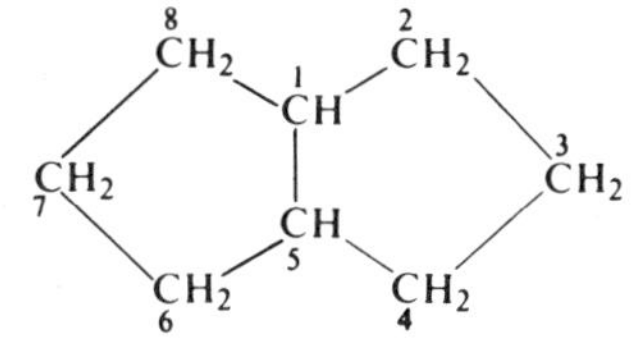 (structure with CH₂ groups numbered 2,3,4,5,6,7,8 and CH at 1)	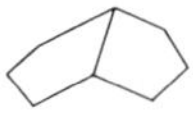	Bicyclo[3.3.0]octane.
9.	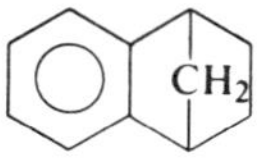 (benzene fused with CH₂ bridge)	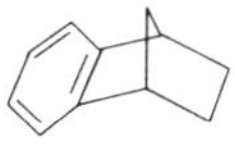	Benzobicyclo[2.2.1]heptene (Benzonorbornene-2)[a]
10.	 (adamantane skeleton)	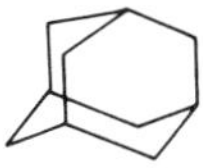	Tricyclo[3.3.1.1^{3,7}]decane (unofficially called adamantane)
11.	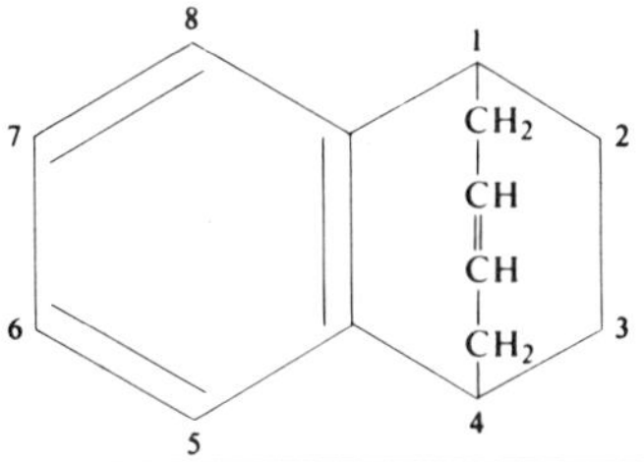 (benzene fused to cyclohexene with CH₂-CH=CH-CH₂ bridge)	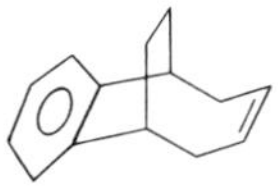	1,2,3,4-Tetrahydro-1,4-(2-buteno)-naphthalene

[a] These are trivial names of special systems that have been retained. See Section H on names of terpenes.

Hydrocarbon systems having three rings are called *tricyclo-*, four rings are called *tetracyclo-*, etc. The number of rings present can be determined by counting the number of cuts necessary to open the cyclic hydrocarbon to a noncyclic hydrocarbon. Thus adamantane is a tricycloalkane since three cuts must be made. The naming of tricyclic systems is similar to that

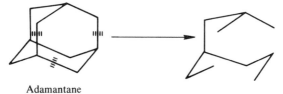

Adamantane

of bicyclic systems except that another bridge must be specified. The IUPAC or CA articles[1,3,6] should be consulted for rules and examples of their use. One example is included in Table 1-6.

The naming of complicated polycyclic systems is often very difficult. Recently, Eckroth[7] has described a very clever systematic procedure for naming polycyclic systems. The student is urged to consult this article if he is faced with the problem of naming a polycyclic system.

Sometimes a new bridge is added on to a parent system. The bridge is

18 STRUCTURE AND NOMENCLATURE OF ORGANIC MOLECULES Chap. 1

named by replacing the *-ane* ending of the bridging molecule with *-ano*, etc. Some of the last compounds in Table 1-6 illustrate this practice.

Many polynuclear aromatic hydrocarbons exist which are really a special kind of bridged compound made up by fusing aromatic rings together. These hydrocarbons often have trivial names and the examples in Table 1-7 illustrate some frequently encountered names and rules. The rules for naming complex fused systems are extensive and the primary references[1,3,6] should be consulted. The student is urged to look at the rules in these primary references in order to get a general idea of how various nomenclature problems are handled.

Table 1-7

IUPAC NAMES OF COMMON POLYCYCLIC AROMATIC COMPOUNDS

	Structure	Name
1.	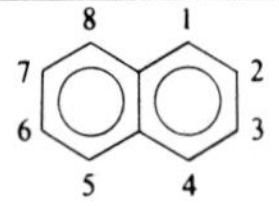	Naphthalene
2.	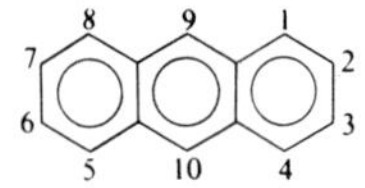	Anthracene
3.	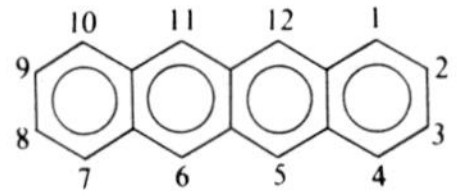	Naphthacene
4.	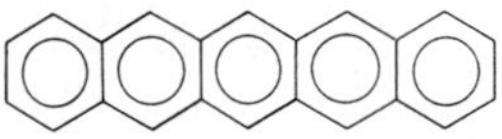	Pentacene
5.	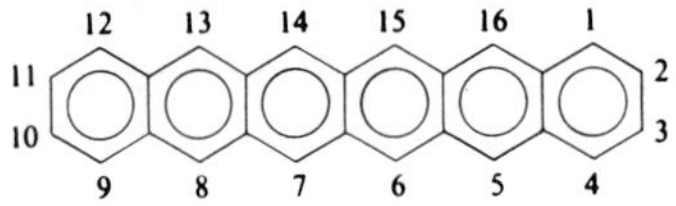	Hexacene
6.	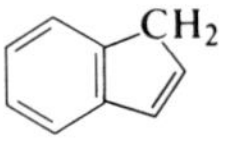	Indene
7.	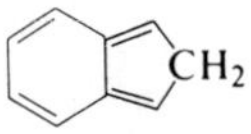	2H-Indene
8.	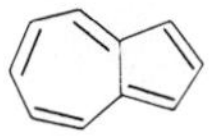	Azulene
9.	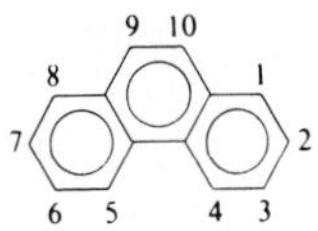	Phenanthrene
10.	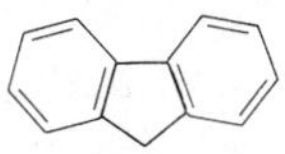	Fluorene
11.	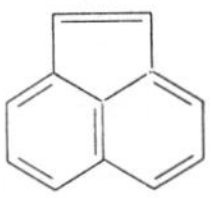	Acenaphthylene

1.4 Nomenclature of Hydrocarbons

Table 1-7 (*continued*)

IUPAC NAMES OF COMMON POLYCYCLIC AROMATIC COMPOUNDS

12.		Rubicene
13.		Benz[a]anthracene
14.		Benzocyclodecene (note: not -pentaene)
15.		1,2,3,4-Tetrahydronaphthalene (unofficially called tetralin)
16.		Perhydronaphthalene (unofficially called decalin)

E. Spiro Hydrocarbons: Two rings that are fused by one carbon atom are called *spiro hydrocarbons* and the joining atom is called a *spiro atom*. These systems are named by preceding the parent name by *spiro-* and inserting numbers in increasing order within brackets that indicate the

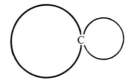

length of the bridges. Compounds with two or more spiro atoms are named by using *dispiro-*, etc., as prefixes and placing the numbers of the bridges in the order in which they come as one starts numbering with a ring atom next to a terminal spiro atom and continues around the system in such a fashion that the spiro atoms receive the lowest numbers possible. The parent name is the name of the open chain hydrocarbon that contains the same number of carbon atoms. Thus, spiro compounds are named in a fashion identical to that for bicyclic compounds except that different prefixes and number orders are used. The examples in Table 1-8 illustrate the various rules.

F. Hydrocarbon Ring Assemblies: *Ring assemblies* are two or more ring systems that are joined to each other by double or single bonds. Biphenyl is a very common ring assembly. As the name "biphenyl" indicates, ring assemblies of identical groups are named by preceding the

Table 1-8
IUPAC NAMES OF SPIRO HYDROCARBONS

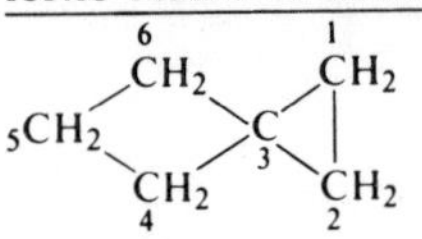

Spiro[2,3]hexane

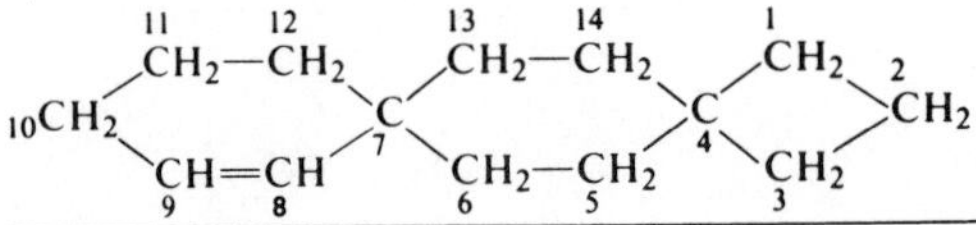

Dispiro[3.2.5.2]tetradec-8-ene

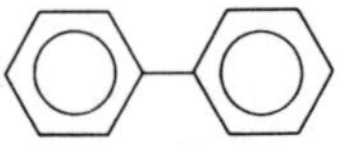

biphenyl

name of the radical with the prefix *bi-*. In some cases, the name of the hydrocarbon may be used.

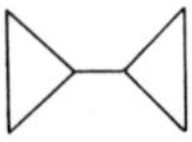

bicyclopropyl or bicyclopropane

For ring assemblies with different groups, the smaller group is named as a substituent of the larger group.

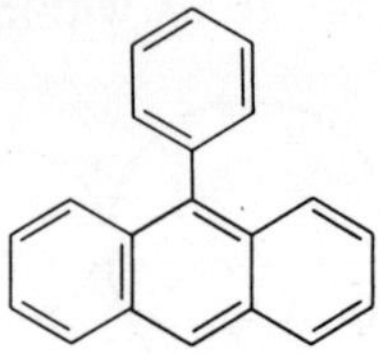

9-phenylanthracene

Other numerical prefixes are used for ring assemblies with three or more groups, such as *ter-* and *quater-* for those with three or four groups, respectively.

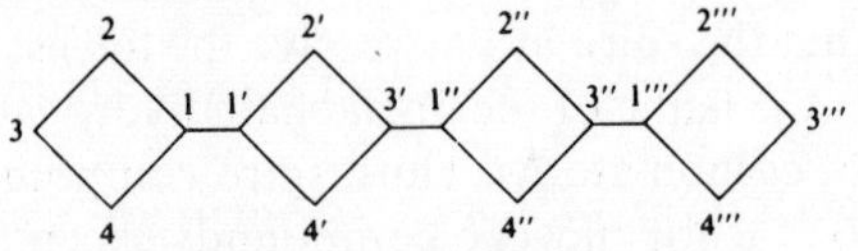

1,1′:3′,1″:3″,1‴-quatercyclobutane

G. Complex Hydrocarbons: Compounds composed of rings and chains are somewhat difficult to name since many choices are available. Thus one could name $C_6H_5-CH_2-C_6H_5$ diphenylmethane, benzylbenzene, or (phenylmethyl)benzene. In general, the name that is the simplest or most informative chemically is used.

1.5 Functional Groups that are Attached to the Skeleton of the Molecule

The base structure is chosen so that it has the maximum number of substituents. Thus **1** is 1-methyl-2-pentylbenzene but **2** is 3-methyl-1-phenylpentane.

1: (benzene ring with CH₃ and CH₂CH₂CH₂CH₂CH₃ substituents)

2: $CH_2CH_2CHCH_2CH_3$ with CH_3 branch, attached to benzene ring

When possible, the smaller unit is named as a substituent. Thus **3** is 1-phenylheptane but **4** is 1-heptylnaphthalene.

3: $CH_3(CH_2)_5CH_2$—(benzene ring)

4: $CH_3(CH_2)_5CH_2$—(naphthalene ring)

H. Terpenes: Many naturally occurring hydrocarbon systems are so common that trivial names have been given to the systems. A selection of these systems is given in Table 1-9. Additional examples can be found in R. K. Hill's book in this series, *Compounds of Nature*.

1.5 FUNCTIONAL GROUPS THAT ARE ATTACHED TO THE SKELETON OF THE MOLECULE

Carbon and hydrogen atoms are not the only kinds of atoms that form strong covalent bonds with carbon. Indeed, oxygen, nitrogen, the halogens, sulfur, phosphorus, and many other elements form strong single and multiple bonds to carbon. For example, the familiar tetraethyl lead, the anti-knock additive in "ethyl gasoline," is a compound containing a lead atom to which four ethyl groups are attached by strong carbon-lead single bonds. Many of the organic compounds that occur in nature contain oxygen and nitrogen. Carbohydrates and fats are almost exclusively composed of carbon, hydrogen, and oxygen. Proteins are mainly made up of these three elements plus nitrogen.

Certain arrangements of atoms of these other elements occur frequently in organic chemistry and these arrangements are the most commonly encountered functional groups. In this section the structure of the most common functional groups will be presented. Little will be said about the physical and chemical properties of these functional groups since these properties will be discussed in Chapters 2 and 3. In this chapter, the student should focus his attention on the structural interrelationships of the various functional groups.

Table 1-9
IUPAC NAMES OF SPECIAL HYDROCARBON SYSTEMS

1.	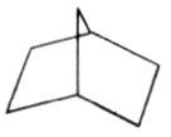	Norbornane
2.	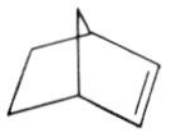	Norbornene
3.		Bornane
4.		p-Menthane
5.	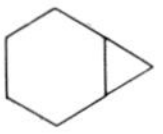	Norcarane
6.		Pinane
7.		Camphane

A. Functional Groups That Contain Carbon-Oxygen Single Bonds: Oxygen is usually divalent, as is apparent from the formula for water, H—O—H. Replacement of one of the hydrogen atoms of water with an alkyl group, R—, leads to *alcohols*, R—O—H, and replacement of both hydrogen atoms by alkyl groups leads to *ethers*, R—O—R. Alcohols constitute a very important class of compounds, as the commonness of wood alcohol, drinking alcohol, and rubbing alcohol suggests. These three alcohols, CH_3OH, C_2H_5OH, and $(CH_3)_2CHOH$, respectively, are actually the three smallest alcohols, but many important alcohols found in nature, such as the steroids, are rather big molecules. Diethyl ether, C_2H_5—O—C_2H_5, is the well-known anesthetic. There is, of course, no reason why both alkyl groups of an ether must be the same.

A modified ether, a three-membered ring formed from two carbon atoms and one oxygen atom, has enough different properties that it is considered a separate functional group called an *epoxide*. Its uniqueness stems, no doubt, from its strain energy, as in the case of cyclopropanes. As the name suggests, "epoxy glues" involve the epoxide functional group.

Acetals,

$$R_2C\begin{matrix}\diagup OR \\ \diagdown OR\end{matrix}$$

form another class of compounds that contain a functional group that is really a modified ether.

1,2-Glycols,

$$\begin{array}{cc} \text{OH} & \text{OH} \\ | & | \\ R_2C - & CR_2 \end{array}$$

are dialcohols that do have some properties not exhibited by most monoalcohols and thus the glycol group can be considered to be a separate functional group.

Just as with water, the hydrogen atoms of hydrogen peroxide, H_2O_2, can be replaced with alkyl groups. *Alkyl hydroperoxides*, R—O—O—H, and *dialkyl peroxides*, R—O—O—R, form the classes of compounds generated.

B. Functional Groups That Contain Nitrogen: The hydrogen atoms of ammonia, NH_3, can be replaced with alkyl groups to give *primary amines*, RNH_2, *secondary amines*, R_2NH, and *tertiary amines*, R_3N. Also, four alkyl groups can be placed around the nitrogen atom to give *quaternary ammonium ions*, R_4N^+, which are, of course, analogous to ammonium ions, NH_4^+. Many important functional groups contain nitrogen-oxygen bonds. Some of the classes of compounds which contain these groups are composed of *nitro* compounds, R—NO_2; *nitroso* compounds, R—N=O; *hydroxyl amines*, R—NH—OH; *alkyl nitrites*, R—O—N=O; *alkyl nitrates*, R—O—NO_2; *amine oxides*, $R_3\overset{+}{N}$—$\overset{-}{O}$; and *oximes*, R_2C=N—O—H. Oximes are important derivatives of aldehydes and ketones, and the oxime group provides our first example of a functional group that contains a carbon-nitrogen double bond. The simplest type of functional group that contains a carbon-nitrogen double bond is found in *imines*, R_2C=N—R. *Nitriles*, R—C≡N, contain carbon-nitrogen triple bonds.

Other important functional groups involve two or more nitrogen atoms. These groups are found in *alkyl hydrazines*, R—NH—NH_2; *hydrazones*, R_2C=N—NH_2; *azo* compounds, R—N=N—R; *azoxy* compounds,

$$\begin{array}{c} O^- \\ | \\ R - \overset{+}{N} = N - R \end{array}$$

carbodiimides, R—N=C=N—R; *diazo* compounds, R_2C=N_2; and *alkyl azides*, R—N_3. Notice the similarities and differences in the structures and names of these functional groups. At this point, the differences may seem slight, but drastic changes in physical and chemical properties may result from these differences.

Compounds that contain a functional group of relatively recent significance are the *nitroxyl radicals*, $R_2\overset{+}{N}-\overset{\cdot\cdot}{\underset{\cdot\cdot}{O}}{:}^{-}$. These compounds are important since they are free radicals that are stable enough to be stored at room temperature.

There are other very important functional groups that contain nitrogen, but these will be discussed in the sections below.

C. Functional Groups That Contain Sulfur: The hydrogen atoms of hydrogen sulfide, H_2S, can be replaced by alkyl groups to give *mercaptans*, R—S—H, and *dialkyl sulfides*, R_2S. Mercaptans have a strong odor; it is the odor associated with bottled or natural gas, since mercaptans are added to the gas so that gas leaks can be readily detected. *n*-Butyl mercaptan, C_4H_9SH, has the odor characteristic of a skunk. Another important sulfur linkage is found in *disulfides*, R—S—S—R.

The three sulfur-containing functional groups mentioned so far are analogous to oxygen-containing functional groups but several important and stable sulfur-containing functional groups have no analogies with oxygen-containing functional groups. Compounds which contain these groups are oxides of sulfur and some of the more common ones are *sulfoxides*,

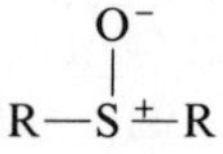

$$R-\overset{\overset{\displaystyle O^-}{|}}{S}^{\pm}-R$$

sulfones

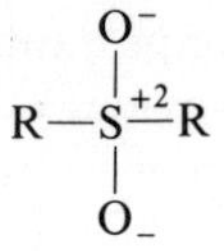

$$R-\overset{\overset{\displaystyle O^-}{|}}{\underset{\underset{\displaystyle O_-}{|}}{S}}{}^{+2}-R$$

sulfinic acids,

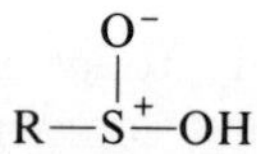

$$R-\overset{\overset{\displaystyle O^-}{|}}{S}{}^{\pm}-OH$$

sulfonic acids,

$$R-\overset{\overset{\displaystyle O^-}{|}}{\underset{\underset{\displaystyle O_-}{|}}{S}}{}^{+2}-OH$$

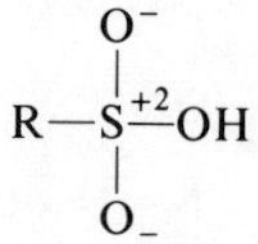

and *sulfonamides*,

$$R-\overset{\overset{\displaystyle O^-}{|}}{\underset{\underset{\displaystyle O_-}{|}}{S}}{}^{+2}-NR_2$$

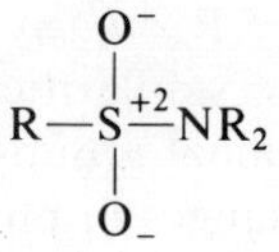

1.5 Functional Groups that are Attached to the Skeleton of the Molecule

Other sulfur-containing functional groups involve the replacement of a carbon-oxygen double bond,

$$\mathrm{\mathop{}\limits^{\diagdown}_{\diagup}C=O}$$

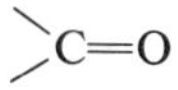

with a carbon-sulfur double bond,

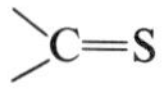

$$\mathrm{\mathop{}\limits^{\diagdown}_{\diagup}C=S}$$

The names of these groups are usually those of the oxygen-containing functional groups with the prefix *thio-* attached. Functional groups that contain carbon-oxygen double bonds are numerous and important and are presented in the next section.

D. Functional Groups That Contain Carbon-Oxygen Double Bonds: The carbon-oxygen double bond,

$$\mathrm{\mathop{}\limits^{\diagdown}_{\diagup}C=O}$$

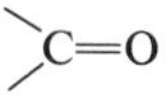

is called a *carbonyl* group and is a part of many important functional groups. Compounds which contain the functional groups that are made up of only a carbonyl group are *aldehydes*,

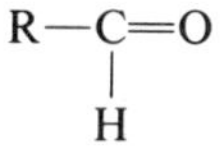

$$\mathrm{R-\underset{H}{\overset{\displaystyle C=O}{|}}}$$

and *ketones*,

$$\mathrm{R-\underset{R}{\overset{\displaystyle C=O}{|}}}$$

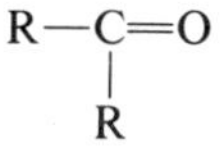

As the formulas indicate, a hydrogen atom and an alkyl group, R, are attached to the carbonyl group in aldehydes, and two alkyl groups are attached to the carbonyl group in ketones. This change leads to several differences in physical and chemical properties.

Many important functional groups are derivatives of *carboxylic acids*,

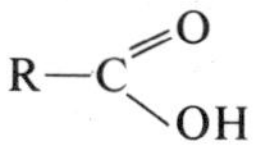

$$\mathrm{R-C\underset{OH}{\overset{\displaystyle O}{\diagup}}}$$

These functional groups have the general formula

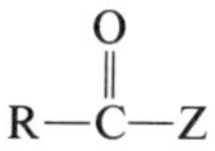

$$\mathrm{R-\overset{\displaystyle O}{\underset{\displaystyle \|}{C}}-Z}$$

where Z is $-OR$, $-NR_2$, $-Cl$, $-Br$, etc. The most important classes of compounds that contain these groups are *esters*, $R-CO-OR$; *amides*, $R-CO-NR_2$; *acid chlorides*, $R-CO-Cl$; *acid bromides*, $R-CO-Br$;

and *acid anhydrides*, R—CO—O—CO—R. The name for the last class of compounds is quite reasonable since acid anhydrides are formed by the removal of one molecule of water from two molecules of a carboxylic acid,

$$\text{R}-\underset{\text{O}}{\overset{\text{O}}{\underset{\|}{\text{C}}}}-\boxed{\text{OH} \quad \text{H}}-\text{O}-\overset{\text{O}}{\underset{\|}{\text{C}}}-\text{R}$$

Nitrogen analogs of acid anhydrides are *imides*, R—CO—NH—CO—R. Two important functional groups which do not have an alkyl group directly attached to the carbonyl group are found in *alkyl ureas*, RNH—CO—NH$_2$, and *urethans*, R—O—CO—NH$_2$. Of course, any or all of the hydrogen atoms shown in these formulas can be replaced with alkyl groups. Three kinds of compounds which contain functional groups that possess the peroxide linkage are important. These classes of compounds are composed of *peroxycarboxylic acids*,

$$\text{R}-\text{C}\begin{matrix}\diagup\text{O}\\ \diagdown\text{O}-\text{O}-\text{H}\end{matrix}$$

peresters,

$$\text{R}-\text{C}\begin{matrix}\diagup\text{O}\\ \diagdown\text{O}-\text{O}-\text{R}\end{matrix}$$

and *diacyl peroxides*, R—CO—O—O—CO—R.

E. Functional Groups That Contain Phosphorus: Many phosphorus-containing compounds are important in nature. Four important types of phosphorus-containing compounds are *trialkylphosphines*, R$_3$P; *trialkylphosphine oxides*, R$_3$P$^\pm$—O$^-$; *trialkyl phosphites*, (RO)$_3$P; and *trialkyl phosphates*, (RO)$_3$PO.

F. Other Important Functional Groups: *Alkyl halides*, R—X, where X is —F, —Cl, —Br, or —I, are important compounds which contain functional groups but do not really fit under the above categories. Alkyl halides are frequently encountered in everyday life. Teflon is a polyfluorocarbon, chloroform is methane with three hydrogen atoms replaced by chlorine atoms, and carbon tetrachloride is methane with all the hydrogen atoms replaced by chlorine atoms. *Iodoso* compounds, R—I=O, contain a stable functional group that consists of an oxide of a halogen.

Acyloins,

$$\text{R}-\underset{\underset{\text{OH}}{|}}{\text{CH}}-\overset{\overset{\text{O}}{\|}}{\text{C}}-\text{R}$$

are really alcohol and ketone combinations, but the proximity of the groups leads to unique properties. The functional group of *ketenes*, $R_2C=C=O$, is a very reactive one. Compounds which contain other common groups are *alkyl isocyanates*, $R—N=C=O$, and *alkyl isocyanides*, $R—\overset{+}{N}\equiv\overset{-}{C}$.

The functional groups that have been presented in this section are the most important functional groups in organic chemistry. There are many other functional groups that are known, but these are of more special interest. Of course, research in organic chemistry is constantly bringing new functional groups to light and showing that either obscure or unknown functional groups have properties that are very important in general.

1.6 NOMENCLATURE FOR ORGANIC COMPOUNDS THAT POSSESS FUNCTIONAL GROUPS CONTAINING CARBON, HYDROGEN, OXYGEN, NITROGEN, HALOGEN, SULFUR, AND PHOSPHORUS ATOMS

The latest official rules for most organic compounds other than hydrocarbons and heterocycles are the *IUPAC 1965 Rules*.[2,4] In this section, the main features of these rules will be discussed, with little attention being paid to selenium and tellurium. These rules involve certain definitions. A *characteristic group* is an atom or group of atoms attached to a compound other than by a direct carbon-carbon bond except for $—C\equiv N$ and $—\overset{|}{C}=Z$ where Z = O, S, Se, Te, NH, or NR. The definition of a characteristic group is more restrictive than that of a functional group since it excludes groups that are composed of only carbon and hydrogen atoms. A *principal group* is the characteristic group which is expressed as the suffix in a particular name. A *substituent* is any atom or group of atoms that replaces a hydrogen atom of a parent compound. *Functional class name* is a name of a class of compounds such as "ketone" or "alcohol" that is used as an ending in radicofunctional nomenclature (defined and discussed below on p. 40). These definitions will become clearer as the reader becomes familiar with the various types of nomenclature.

There are several types of nomenclature used to name compounds containing various characteristic groups. The particular type of nomenclature that one uses will depend on several things. Certain relationships between members of a group of compounds will often be best indicated by a certain type of nomenclature. Many times one type of nomenclature leads to a much simpler name. Often only one type of nomenclature can be used to name a certain compound. The student should be familiar with all the types of nomenclature since they are all encountered in the literature.

Table 1-10
SUBSTITUTIVE NOMENCLATURE OF CHARACTERISTIC GROUPS

Characteristic group	Structure of group	Prefix	Suffix	Structure of example	Name of example
Alcohols (and phenols)	—OH	hydroxy-	-ol	CH_3CH_2OH	Ethanol
Aldehydes	—CH=O —[C]H=O[b]	formyl- oxo-	-carbaldehyde[a] -al	$CH_3CH=O$ $CH_3CH=O$	Methanecarbaldehyde Ethanal
Ketones	R—[C(=O)]—R	oxo-	-one	$CH_3CH_2\overset{O}{\underset{\parallel}{C}}CH_2CH_3$	3-Pentanone
Carboxylic acids	—C(=O)—OH	carboxy-	-carboxylic acid	CH_3COOH	Methanecarboxylic acid
	[C(=O)]—OH	—	-oic acid	CH_3COOH	Acetic acid; Ethanoic acid[c]
Thiols	—SH	mercapto-	-thiol	CH_3CH_2SH	Ethanethiol
Ethers	—O—R	alkoxy-	—	$CH_3CH_2OCH_2CH_3$	Ethoxyethane
Epoxide	(epoxide ring)	epoxy-	—	$CH_2\underset{O}{\overset{}{-}}CH_2$	Epoxyethane
Sulfides	—S—R	alkylthio-	—	$CH_3CH_2SCH_2CH_3$	Ethylthioethane
Disulfides	—S—SR	alkyldithio-	—	$CH_3CH_2SSCH_2CH_3$	Ethyldithioethane
Sulfoxides	—SO—R	alkylsulfinyl-	—	CH_3CH_2—SO—CH_2CH_3	Ethylsulfinylethane
Sulfones	—SO_2—R	alkylsulfonyl-	—	CH_3CH_2—SO_2—CH_2CH_3	Ethylsulfonylethane

1.6 Nomenclature for Organic Compounds

Acetals[d]	$>C<^{OR}_{OR}$	1,1-dialkoxy-	—	$CH_3CH(OCH_2CH_3)_2$	1,1-Diethoxyethane
Isocyanates	—NCO	isocyanato-	—	CH_3CH_2NCO	Isocyanatoethane
Azo compounds	R—N=N—R R—N=N—R'	azo- azo-	— —	$CH_3CH_2N_2CH_2CH_3$ $CH_3CH_2N_2CH_3$	Azoethane Ethaneazomethane
Hydrazines	$R_2N—NR_2$	hydrazino-	-hydrazine	$[(CH_3CH_2)_2N]_2$	Tetraethylhydrazine
Sulfonic acids	—SO_3H	sulfo-	-sulfonic acid	$CH_3CH_2SO_3H$	Ethanesulfonic acid
Sulfonamides	—SO_2—N$<^{\|}_{\|}$	-sulfonamido- -sulfonylamino-	-sulfonamide	$CH_3CH_2SO_2NH_2$ $CH_3CH_2SO_2—NH—CH_2COOH$	Ethanesulfonamide Ethanesulfonamidoacetic acid or ethylsulfonylaminoacetic acid
Sulfinic acids	R—SO_2H	sulfino-	-sulfinic	$CH_3CH_2SO_2H$	Ethanesulfinic acid
Amines	—NH_2	amino-	-amine	$(CH_3CH_2)_3N$	Triethylamine
Imines	=NH	-imino-	-imine	$CH_3CH=NH$	Ethylimine
Ureas	RN—C—NR $\|$ $\|\|$ $\|$ H O H	ureido-	-urea	$(CH_3CH_2NH)_2C=O$	N,N'-Diethylurea or 1,3-diethylurea
Azides	—N_3	azido-	—	$CH_3CH_2N_3$	Azidoethane
Halides	—X (X = Br, Cl, F, I)	bromo-, chloro-, fluoro-, iodo-	—	CH_3CH_2Cl	Chloroethane
Iodoso compounds	—IO	iodosyl-	—	CH_3CH_2IO	Iodosylethane

Table 1-10 (continued)
SUBSTITUTIVE NOMENCLATURE OF CHARACTERISTIC GROUPS

Characteristic group	Structure of group	Prefix	Suffix	Structure of example	Name of example
Nitro compounds	$-NO_2$	nitro-	—	$CH_3CH_2NO_2$	Nitroethane
Nitroso compounds	$-NO$	nitroso-	—	CH_3CH_2NO	Nitrosoethane
Diazo compounds	$=N_2$	diazo-	—	$CH_3CH=N_2$	Diazoethane
Peroxides	$-O-O-R$	alkyldioxy-	—		Ethyldioxyethane
Acid salt	$-COO^-M^+$ $-[C]OO^-M^+$	— —	-carboxylate -oate	CH_3COONa CH_3COONa	Sodium methanecarboxylate Sodium acetate or sodium ethanoate
Ester	$-COOR$ $-[C]OOR$	alkoxycarbonyl- —	-carboxylate -ate	CH_3COOCH_3 CH_3COOCH_3	Methyl methanecarboxylate Methyl acetate or methyl ethanoate
Amide	$-CON{\Big\langle}$ $-[C]ON{\Big\langle}$	carbamoyl- —	-carboxamide -amide	CH_3CONH_2 CH_3CONH_2 $CH_3CH_2CON(CH_3)_2$	Methanecarboxamide Acetamide or ethanamide N,N-Dimethylpropanamide or N,N-dimethylpropionamide
Nitrile	$-C{\equiv}N$ $-[C]{\equiv}N$	cyano- —	-carbonitrile -nitrile	CH_3CN CH_3CN	Methanecarbonitrile Acetonitrile or ethanenitrile
Acid halide	$-CO-X$ (X = halide) $-[C]O-X$	haloformyl- —	-carbonyl halide -oyl halide	CH_3COCl CH_3COCl	Methanecarbonyl chloride Acetyl chloride or ethanoyl chloride

1.6 Nomenclature for Organic Compounds

Class	Structure	Suffix	Prefix	Example	Name
Acid anhydrides	—CO—O—CO—	-carboxylic -anhydride	—	$(CH_3CO)_2O$	Methanecarboxylic anhydride
	—[C]O—O—[C]O—	-oic anhydride	—	$(CH_3CO)_2O$	Acetic anhydride or ethanoic anhydride
				(succinic anhydride structure)	Succinic anhydride
Imide	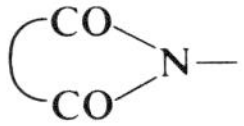 CO–N–CO (ring)	—	-carboximide	(1,3-propanedicarboximide structure, NH in six-membered ring with two C=O)	1,3-Propanedicarboximide
	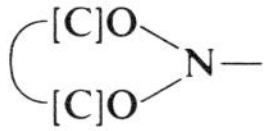 [C]O–N–[C]O (ring)	—	-imide	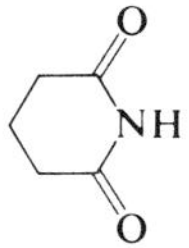 (glutarimide structure)	Glutarimide
Oxime	$\rangle$[C]=NOH	-oxime (after name of aldehyde or ketone)	hydroxyimido-	$CH_3CH_2\underset{\underset{NOH}{\parallel}}{C}CH_2CH_3$	3-Pentanone oxime
				$HC(=NOH)-CH_2CH_2COOH$	4-Hydroxyimidobutyric acid
Hydroxyl-amine	—NH—OH	-hydroxylamine	hydroxyamino-	CH_3CH_2NHOH	N-Ethylhydroxylamine
				$H_2\overset{HNOH}{\underset{\mid}{C}}CH_2CH_2COOH$	4-Hydroxyaminobutyric acid

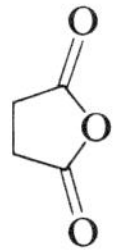

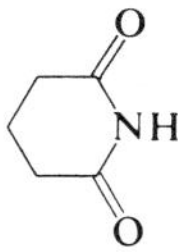

Table 1-10 *(continued)*
SUBSTITUTIVE NOMENCLATURE OF CHARACTERISTIC GROUPS

Characteristic group	Structure of group	Prefix	Suffix	Structure of example	Name of example
Amine oxides	$-\overset{\mid}{\underset{\mid}{N}}{}^{+}-O^{-}$	—	-oxide (after name of the amine)	$(CH_3CH_2)_3NO$	Triethylamine oxide
Carbodiimides	$-N=C=N-$	—	-carbodiimide	$CH_3CH_2-N=C=N-CH_2CH_3$	Diethylcarbodiimide
Acyloins	$R-[C]H-\underset{\mid}{\overset{OH}{[C]}}-R$	—	-oin (in place of -ic acid or -oic acid of the trivial acid RCOOH)	$CH_3CH-\underset{\mid}{\overset{OH}{C}}-CH_3$ $\overset{\parallel}{O}$	Acetoin
Peroxy acids	$-\overset{O}{\overset{\parallel}{C}}-O-O-H$	—	-peroxycarboxylic	CH_3COOOH	Methaneperoxycarboxylic acid
	$=[\overset{\parallel}{O}]-O-O-H$	—	peroxy—oic acid	CH_3COOOH $CH_3(CH_2)_6COOOH$	Peracetic acid[e] Peroxyoctanoic acid

[a] *-Carboxaldehyde* is the suffix that has been used and is still often encountered.
[b] The carbon enclosed in brackets is included in the name of the parent compound.
[c] The trivial name is preferred.
[d] The name "ketal" is no longer used.
[e] The trivial names for performic, peracetic, and perbenzoic acids have been accepted.

1.6 Nomenclature for Organic Compounds

A. Substitutive Nomenclature: The most important and systematic type of nomenclature is *substitutive nomenclature*. As the name implies, the characteristic group is named as a substituent and the appropriate prefix or suffix is added to the name of the parent hydrocarbon. Some characteristic groups can be cited as suffixes or prefixes but some can only be cited as prefixes. One must be careful to name the parent hydrocarbon correctly. For example, "*tert*-butanol," a name frequently encountered, is an incorrect name, since *tert*-butane does not exist. Substitutive nomenclature for the most frequently encountered functional groups is presented in Table 1-10.

The acids and their derivatives deserve special attention. The IUPAC has agreed to retain many of the trivial names of acids since they are so commonly used. The Geneva (or systematic) nomenclature considers acids derived from the corresponding hydrocarbons and the name is generated by replacing the *-e* with *-oic acid*. Thus CH_3COOH is properly called either "acetic acid" or "ethanoic acid." In Table 1-11 are listed the most important trivial names that are accepted and preferred by IUPAC for carboxylic acids and their acyl radicals.

Table 1-11

TRIVIAL NAMES OF CARBOXYLIC ACIDS AND THEIR ACYL RADICALS ACCEPTED AND PREFERRED BY IUPAC

Number of carbon atoms	Structure	Name of acid	Name of acyl radical
1	HCOOH	Formic	Formyl
2	CH_3COOH	Acetic	Acetyl
3	CH_3CH_2COOH	Propionic	Propionyl
4	$CH_3(CH_2)_2COOH$	Butyric	Butyryl
5	$CH_3(CH_2)_3COOH$	Valeric	Valeryl
12	$CH_3(CH_2)_{10}COOH$	Lauric[a]	Lauroyl[a]
14	$CH_3(CH_2)_{12}COOH$	Myristic[a]	Myristoyl[a]
16	$CH_3(CH_2)_{14}COOH$	Palmitic[a]	Palmitoyl[a]
18	$CH_3(CH_2)_{16}COOH$	Stearic[a]	Stearoyl[a]
2	HOOCCOOH	Oxalic	Oxalyl[b]
3	$HOOCCH_2COOH$	Malonic	Malonyl[b]
4	$HOOC(CH_2)_2COOH$	Succinic	Succinyl[b]
5	$HOOC(CH_2)_3COOH$	Glutaric	Glutaryl[b]
6	$HOOC(CH_2)_4COOH$	Adipic[a]	Adipoyl[b]
	$CH_2=CHCOOH$	Acrylic	Acryloyl
	$CH_2=C(CH_3)COOH$	Methacrylic	Methacryloyl
	$CH_3CH=CHCOOH$	Crotonic (*trans*)	Crotonoyl
	(phenyl)-COOH	Benzoic	Benzoyl
	(naphthyl)-COOH	Naphthoic	Naphthoyl

Table 1-11 (*continued*)

TRIVIAL NAMES OF CARBOXYLIC ACIDS AND THEIR ACYL RADICALS ACCEPTED AND PREFERRED BY IUPAC

Number of carbon atoms	Structure	Name of acid	Name of acyl radical
	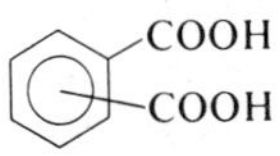 C₆H₄(COOH)₂ (ortho, meta, para)	1,2 is phthalic 1,3 is isophthalic 1,4 is terephthalic	Phthaloyl[b] Isophthaloyl[b] Terephthaloyl[b]
	HCCOOH ∥ HCCOOH	Maleic	Maleoyl[b]
	HCCOOH ∥ HOOCCH	Fumaric	Fumaroyl[b]

[a] These names are recommended for unsubstituted acids only.
[b] These names are for diradicals, —CO—X—CO—.

Certain special arrangements of characteristic groups have been given special names. Three such important groups are listed in Table 1-12.

Table 1-12

ARRANGEMENTS OF CHARACTERISTIC GROUPS THAT POSSESS SPECIAL NAMES

Structure	Name
H₂C=C=O	Ketene
H₂N—C(=NH)—NH₂	Guanidine
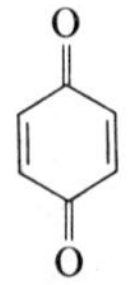 O=C₆H₄=O (para)	*p*-Benzoquinone

If a compound contains more than one characteristic group, then only one group, the principal group, can be denoted by a suffix. The priority of a group decreases in this order: cations, acids, derivatives of acids, nitriles, aldehydes, ketones, alcohols, hydroperoxides, and amines. All characteristic groups except the principal group are then named by prefixes. Many rules are given for selection of the principal chain. One of the most important rules is that the principal chain should contain the maximum number of principal groups. For relatively simple compounds, the selection of the principal chain is no problem. For more complex compounds, reference to the primary articles[2,4] should be made. After the principal chain is selected, it is numbered according to the rules for hydrocarbons.

1.6 Nomenclature for Organic Compounds

If these rules leave a choice, one must refer to another set of rules which leads to one preferred system. Again, these rules are too involved to cite here in their entirety, but a very important rule is that the principal group which is cited as a suffix should receive the lowest possible number. The characteristic groups other than the principal group are then cited by prefixes arranged in alphabetical order preceded by the appropriate number. These and other rules are illustrated by the examples in Table 1-13. Notice that, in general, the final -*e* of parent compounds is dropped when followed by a suffix that begins with "a," "i," or "o." Also the final -*a* of a multiplying affix such as *tetra-* is dropped when followed by a suffix that begins with an "a" or "o."

Table 1-13
SUBSTITUTIVE NOMENCLATURE OF VARIOUS COMPOUNDS THAT CONTAIN CHARACTERISTIC GROUPS

1.
$$\underset{\underset{\text{OH}}{|}}{\overset{1\ \ \ 2\ \ \ 3\ \ \ 4\ \ \overset{\text{CH}_3}{\overset{|}{5}}\ \ \ 6\ \ \ 7}{\text{CH}_3\text{CHCH}_2\text{CH}_2\text{CHCH}_2\text{CH}_3}}$$

5-Methyl-2-heptanol

2.
$$\underset{\underset{\text{OH}}{|}}{\overset{6\ \ \ 5\ \ \ 4\ \ \ 3\ \ \overset{\overset{1}{\text{CH}_2\text{OH}}}{\overset{|}{2}}}{\text{CH}_3\text{CHCH}_2\text{CH}_2\text{CHCH}_2\text{CH}_3}}$$

2-Ethyl-1,5-hexanediol

3.
$$\underset{\underset{\text{OH}}{|}\ \ \ \ \underset{\underset{\text{CH}_3}{|}}{\underset{|}{\text{CH}_3\text{COH}}}}{\overset{\overset{\text{CH}_3\text{CHOH}}{|}}{\text{CH}_3\text{CHCH}_2\text{CH}_2\text{CCH}_2\text{CH}_3}}$$

3-Ethyl-3-(1-hydroxy-ethyl)-2-methyl-2,6-heptanediol

4.
$$\underset{\underset{\text{Cl}}{|}}{\text{CH}_3\text{CHCH}_2\text{CH}_3}$$

2-Chlorobutane

5.
$$\underset{\underset{\text{Br}}{|}\ \ \ \ \ \underset{\text{F}}{|}}{\text{CH}_3\text{CH}_2\text{CHCH}_2\text{CH}_2\text{CHCH}_3}$$

5-Bromo-2-fluoroheptane

6. $\text{CH}_3\text{CH}_2\text{CH}_2\text{NHCH}_2\text{CH}_3$

N-Ethylpropylamine

7.
$$\underset{\underset{\text{NH}_2}{|}}{\text{CH}_3\text{CH}_2\text{CHCH}_2\text{CH}_3}$$

3-Pentylamine

8.
$$\underset{\underset{\text{OH}}{|}\ \ \ \ \ \ \underset{\text{CH}_3}{|}\ \ \ \ \ \ \ \ \ \ \underset{\text{Cl}}{|}}{\overset{8\ \ \ \ \ 7\overset{\diamond}{\ }\ \ 6\ \ 5\ \ \ 4\ \ \ \overset{\overset{\text{O}}{\|}}{3}\ \ 2\ \ \ 1}{\text{CH}_2-\text{CH}-\text{C}=\text{CH}-\text{CH}_2-\text{C}-\text{CH}-\text{CH}_3}}$$

2-Chloro-7-cyclobutyl-8-hydroxy-6-methyl-5-octen-3-one

Table 1-13 (*continued*)

SUBSTITUTIVE NOMENCLATURE OF VARIOUS COMPOUNDS THAT CONTAIN CHARACTERISTIC GROUPS

#	Structure	Name
9.	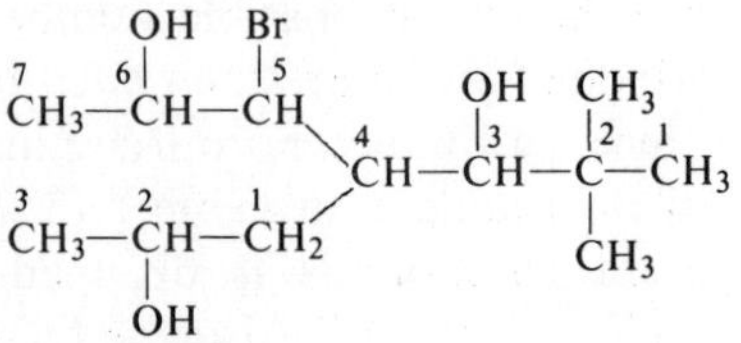 $\overset{7}{C}H_3-\overset{6}{C}H(OH)-\overset{5}{C}H(Br)-\overset{4}{C}H(-CH(OH)CH_2\overset{3}{C}H(CH_3)\overset{2}{C}H_3)$	5-Bromo-4-(2-hydroxypropyl)-2,2-dimethyl-3,6-heptanediol
10.	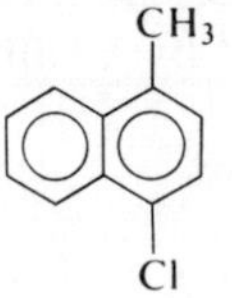 1-chloro-4-methylnaphthalene structure	1-Chloro-4-methylnaphthalene
11.	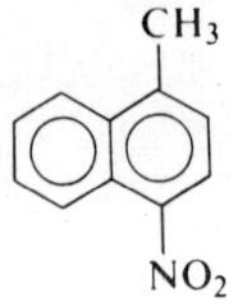 1-methyl-4-nitronaphthalene structure	1-Methyl-4-nitronaphthalene
12.	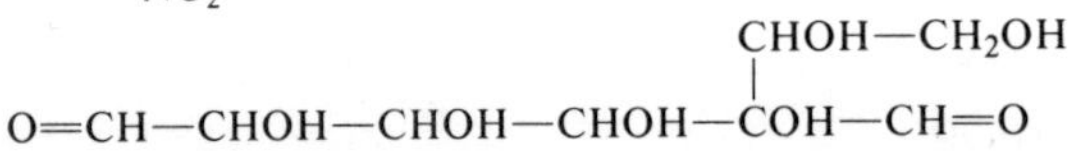 $O=CH-CHOH-CHOH-CHOH-\underset{\underset{CHOH-CH_2OH}{\mid}}{COH}-CH=O$	2-(1,2-Dihydroxyethyl)-2,3,4,5-tetrahydroxyhexanedial
13.	$\overset{5}{C}H_3\overset{4}{C}H_2\overset{3}{C}H_2\overset{2}{C}H_2\overset{1}{C}O_2H$ $\quad\overset{4}{\ }\overset{3}{\ }\overset{2}{\ }\overset{1}{\ }$	Pentanoic acid, valeric acid, or 1-butanecarboxylic acid
14.	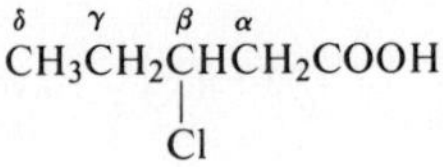 $\overset{\delta}{C}H_3\overset{\gamma}{C}H_2\overset{\beta}{C}H(Cl)\overset{\alpha}{C}H_2COOH$	3-Chloropentanoic acid, 3-chlorovaleric acid, or β-chlorovaleric acid,[a] or 2-chloro-1-butanecarboxylic acid
15.	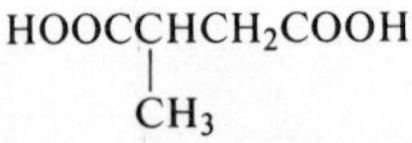 $HOOCCH(CH_3)CH_2COOH$	2-Methylbutanedioic acid or 2-methylsuccinic acid
16.	$CH_2=CHCH_2CH_2CH_2COOH$	5-Hexenoic acid
17.	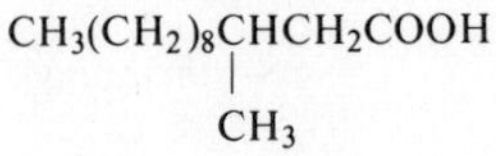 $CH_3(CH_2)_8CH(CH_3)CH_2COOH$	3-Methyldodecanoic acid (not 3-methyllauric acid)[b]
18.	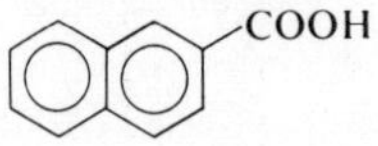 2-naphthoic acid structure (COOH)	2-Naphthalenecarboxylic acid or 2-naphthoic acid
19.	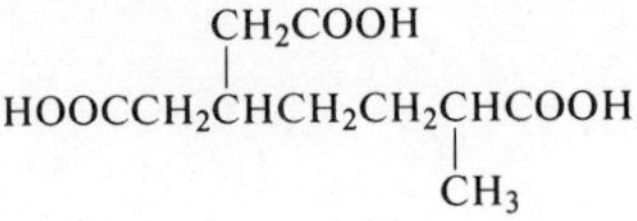 $HOOCCH(CH_2COOH)CH_2CH_2CH(CH_3)COOH$	5-Carboxymethyl-2-methylheptanedioic acid

1.6 Nomenclature for Organic Compounds

Table 1-13 (*continued*)
SUBSTITUTIVE NOMENCLATURE OF VARIOUS COMPOUNDS THAT CONTAIN CHARACTERISTIC GROUPS

20.	(cyclopentane ring)—COOH	Cyclopentanecarboxylic acid
21.	(norbornane ring)—COOH	1-Norbornanecarboxylic acid
22.	HOOC—CH(CH$_2$)(CH$_2$)CH—COOH	1,3-Cyclobutanedicarboxylic acid
23.	CH$_3$CHOHCHOHCH$_2$CH$_2$CHOHCHOHCH$_3$	2,3,6,7-Octanetetrol

[a] With trivial names, Greek letters are often used to denote the position of a substituent. Note that position α corresponds to position 2.
[b] See footnote a to Table 1-11.

B. Nomenclature for Lactones, Lactams, Sultones, and Sultams: Lactones are cyclic esters and lactams are cyclic amides. Lactones are named by adding the suffix *-olide* or *-carbolactone* to the name of the corresponding hydrocarbon using numbers to show the position of ring closure. Lactams are named in a similar fashion except that the suffix *-lactam* is used. Sultones and sultams are analogous to lactones and lactams and have an —SO$_2$— in place of the —CO—. These compounds may also be named as heterocycles and certain ones have acceptable trivial names. The examples in Table 1-14 will clarify these practices.

C. Nomenclature for Free Radicals, Ions, and Radical Ions: Free radicals are given the name which is used when the radical is used as a substituent. When the name of the radical ends in *-y*, the ending is changed to *-yl*. For oxygen radicals, such as CH$_3$O·, it is common to use the ending *-y*, not *-yl*. The unofficial but more common nomenclature for CH$_3$O·, then, is *methoxy*. Compounds having the structure Z$_2$C: are called *carbenes*. Examples are found in Table 1-15.

A compound that contains a positive carbon atom, a carbonium ion, is named by (1) addition of the word *cation* after the name of the corresponding radical or (2) addition of the suffix *-ium* to the name of the corresponding radical. Compounds that contain other elements that are cations are named in a similar fashion. Thus trivalent oxygen compounds are *oxonium* compounds, trivalent sulfur compounds are *sulfonium* compounds, divalent halogen compounds are *halonium* ions, etc. These rules are illustrated in Table 1-16. Diazonium compounds, RN$_2^+$X$^-$, are named

Table 1-14

NOMENCLATURE FOR LACTONES AND LACTAMS

#	Structure	Name
1.	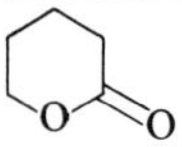 (six-membered ring with O and C=O)	5-Pentanolide or δ-valerolactone or 4-butanecarbolactone
2.	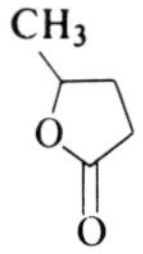 (five-membered ring with O, C=O, and CH₃ substituent)	4-Pentanolide or γ-valerolactone or 3-butanecarbolactone
3.	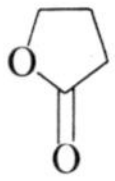 (five-membered ring with O and C=O)	4-Butanolide or γ-butyrolactone or 3-propanecarbolactone
4.	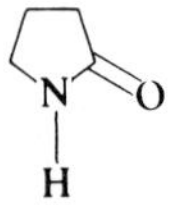 (five-membered ring with NH and C=O)	4-Butanelactam or γ-butyrolactam
5.	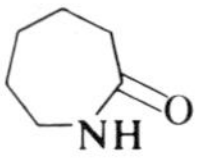 (seven-membered ring with NH and C=O)	6-Hexanelactam or ε-caprolactam
6.	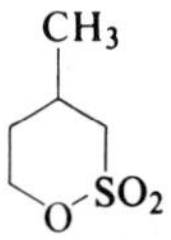 (six-membered ring with O, SO₂, and CH₃ substituent)	2-Methyl-1,4-butanesultone
7.	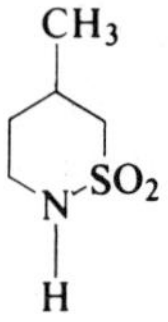 (six-membered ring with NH, SO₂, and CH₃ substituent)	2-Methyl-1,4-butanesultam
8.	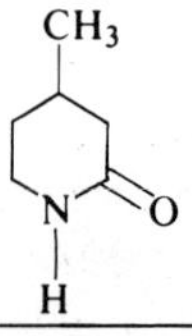 (six-membered ring with NH, C=O, and CH₃ substituent)	3-Methyl-5-pentanelactam

Table 1-15

NOMENCLATURE OF FREE RADICALS

Ph₃C·	Triphenylmethyl or trityl
CH₃O·	Methoxyl (unofficially called Methoxy)
(CH₃)₃C—O—O·	*tert*-Butylperoxyl (unofficially called *t*-Butylperoxy)
Cl₂C:	Dichlorocarbene or dichloromethylene

1.6 Nomenclature for Organic Compounds

by adding the suffix -*diazonium* to the name of the parent compound RH with the name of the anion X being placed afterward.

Table 1-16
NOMENCLATURE OF CATIONS

1.	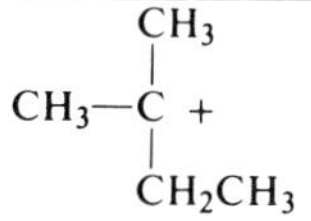 $\underset{\underset{CH_2CH_3}{\mid}}{\overset{\overset{CH_3}{\mid}}{CH_3-C\,+}}$	1,1-Dimethylpropyl cation or 1,1-dimethylpropylium
2.	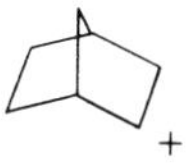 (norbornyl structure) +	Norbornyl cation or norbornylium
3.	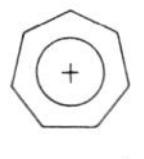 (cycloheptatrienyl ring with +)	Cycloheptatrienylium or tropylium
4.	Ph_4N^+	Tetraphenylammonium
5.	$(CH_3CH_2)_3O^+$	Triethyloxonium
6.	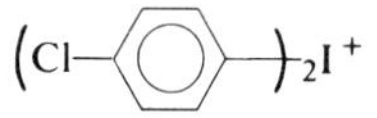 $(Cl-C_6H_4-)_2I^+$	Di-*p*-chlorophenyliodonium
7.	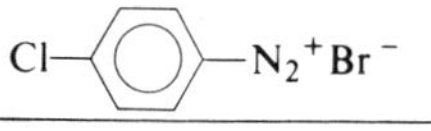 $Cl-C_6H_4-N_2^+Br^-$	*p*-Chlorobenzenediazonium bromide

Anions derived from alcohols are named by adding -*olate* or -*oxide* to the name of the alkyl portion. Carbon anions, carbanions, are usually named by using the suffix -*ide*. Names of several anions are given in Table 1-17.

Table 1-17
NOMENCLATURE OF ANIONS

CH_3CH_2ONa	Sodium ethanolate or sodium ethoxide
$CH_3CH_2CH_2CH_2OK$	Potassium butanolate or potassium butoxide
$HC\equiv CAg$	Silver acetylide
$(C_6H_5)_3CK$	Potassium triphenylmethylide

Radical cations are named by (1) adding the word *cation* to the name of the compound which has the same molecular formula or (2) by adding the suffix -*yl* to the name of the cation. Radical anions are named by (1) adding the word *anion* to the name of the compound which has the same molecular formula or (2) by adding the suffix -*ide* to the name of the corresponding radical formed by the addition of a proton to the radical anion. The names of ketyls and semiquinones are retained. These practices are illustrated in Table 1-18.

Table 1-18

NOMENCLATURE OF RADICAL CATIONS AND RADICAL ANIONS

1.	[benzene ring with +·] or $C_6H_6^{+\cdot}$		Benzene cation or benzeniumyl
2.	[pyridine with N$_+^{\cdot}$]		Pyridine 1-cation or 1-pyridiniumyl
3.	$\begin{bmatrix}\text{NO}_2\\\text{[ring]}\end{bmatrix}^{\cdot-}$ or $\begin{bmatrix}\text{NO}_2\\\text{[ring]}^-\end{bmatrix} \leftrightarrow \text{etc.}$ or $C_6H_5NO_2^{\cdot-}$		Nitrodihydrophenylide or nitrobenzene anion
4.	[Ph$_2$C(O$^-$)· Na$^+$]		Sodium diphenylketyl
5.	$^-\!:\!\ddot{O}\!-\!\!\!\bigcirc\!\!\!-\!\dot{O}\!:$		p-Benzosemiquinone anion

D. Radicofunctional Nomenclature: Substitutive nomenclature is usually the preferred type of nomenclature. However, radicofunctional nomenclature is used often, and frequently permits one to emphasize the functional group of the molecule. In radicofunctional nomenclature, the functional class name indicates the presence of the characteristic group. The functional class name is preceded by the names of the radicals that are attached to the characteristic group. In Table 1-19 are presented the functional class names of the most common groups. The groups are listed in decreasing priority. Examples of the use of radicofunctional names are found in Table 1-20.

E. Additive Nomenclature: Additive nomenclature is often used to name molecules that can be considered to be made up of a parent molecule to which atoms have been added. As the name implies, additive nomenclature consists of adding the names of these additional atoms to the name of the parent system. Most commonly, additive nomenclature is used when hydrogens are added to a structure. Also, the prefix *homo-* is used to denote the addition of a —CH$_2$— into a ring. Other uses of this type of nomenclature involve the addition of the name of the added atom or atoms after the name of the parent compound. Examples of additive nomenclature are found in Table 1-21.

F. Subtractive Nomenclature: Subtractive nomenclature is the opposite of additive nomenclature. Atoms that have been removed from some parent structure are denoted by prefixes. The loss of hydrogens from a compound that possesses a trivial name is often indicated by the prefix *dehydro-*. The prefix *nor-* has sometimes been used to indicate that all the

1.6 Nomenclature for Organic Compounds

Table 1-19

FUNCTIONAL CLASS NAMES USED IN RADICOFUNCTIONAL NOMENCLATURE

Group	Functional class name
Halide in acid halides, $R-\underset{\underset{O}{\|\|}}{C}-X$	Name of halide
$-CN$	Cyanide
$-NCO$	Isocyanate
$-CO-$	Ketone
$-OH$	Alcohol
$-O-OH$	Hydroperoxide
$-O-O-$	Peroxide
$-O-$	Ether or oxide
$-S-$	Sulfide
$-SO-$	Sulfoxide
$-SO_2-$	Sulfone
$-X$	Halide
$-N_3$	Azide

Table 1-20

RADICOFUNCTIONAL NOMENCLATURE OF VARIOUS COMPOUNDS

1.	$CH_3CH_2CH_2\underset{\underset{O}{\|\|}}{C}-Cl$	Butyryl chloride
2.	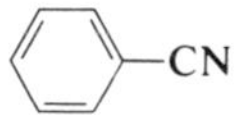 C_6H_5-CN	Phenyl cyanide
3.	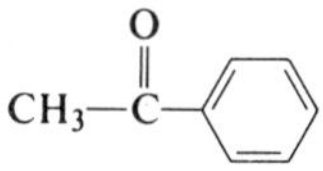 $CH_3-\underset{\underset{O}{\|\|}}{C}-C_6H_5$	Methyl phenyl ketone
4.	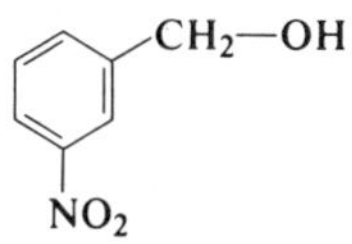 m-$O_2N-C_6H_4-CH_2-OH$	m-Nitrobenzyl alcohol
5.	$CH_3-SO-CH_3$	Dimethyl sulfoxide

methyl groups of a cyclic system have been replaced with hydrogens. More generally, *nor-* is used to indicate the removal of a $-CH_2-$ group, especially in steroids. If the lost $-CH_2-$ was part of a chain, its position is indicated by a number; if it was part of a ring, its position is indicated by a capital letter. The prefix *de-* placed before any group of atoms indi-

Table 1-21

ADDITIVE NOMENCLATURE

Structure	Name
1,4-Dihydrobenzene (cyclohexadiene with H,H at 1 and 4 positions)	1,4-Dihydrobenzene
Cyclooctatetraene cation (+)	Homotropylium
$CH_3-CH-CH-CH_3$ with bridging O	2-Butene oxide
$CH_3-CHCl-CHCl-CH_3$	2-Butene dichloride

cates the replacement of that group with hydrogen. In Table 1-22, several illustrations of subtractive nomenclature are presented.

G. Conjunctive Nomenclature: Conjunctive nomenclature was formerly called "additive nomenclature" by *Chemical Abstracts* where it is frequently used in inverted indexes. Conjunctive nomenclature is used when a cyclic component is attached to an acyclic component. Each component is named as if it were a separate compound. The two names are then joined together with the cyclic name first. It is understood that each component loses one hydrogen atom at the position where it is joined to the other component. Numerals are used to indicate the point of attachment to the cyclic component and Greek letters are used to indicate the point of attachment of substituents to the side chain. The side chain is considered to extend from the principal group to the ring. These and other rules are illustrated in Table 1-23.

H. Replacement Nomenclature: Replacement nomenclature is an extension to both cyclic and acyclic systems of "a" nomenclature that is used for heterocycles. It is most commonly used for acyclic systems when there are several hetero atoms that consequently make other systems of nomenclature difficult to apply. In replacement nomenclature the compound is named as if it were a hydrocarbon and the positions and nature of the hetero atoms present are indicated by prefixes. The prefixes for the three most common hetero atoms, oxygen, sulfur, and nitrogen, are *oxa-*, *thia-*, and *aza-*. These prefixes are listed in decreasing priority. Examples of replacement nomenclature are found in Table 1-24.

I. Nomenclature of Phosphorus Compounds: The IUPAC has not yet published definitive rules for naming phosphorus compounds. However, the *Chemical Abstracts* article[6] does give the present practices based on an article by the American Chemical Society in *Chemical and Engineering News*.[8] The examples in Table 1-25 illustrate these rules. The *Chemical and*

1.7 Nomenclature for Cis-Trans or Geometric Isomers

Table 1-22

SUBTRACTIVE NOMENCLATURE

1.	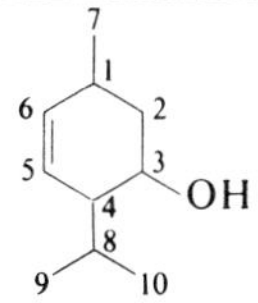 (structure)	5,6-Didehydromenthol[a]
2.	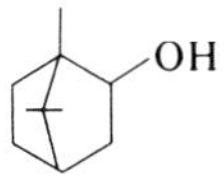 (structure)	Borneol
3.	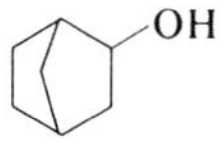 (structure)	Norborneol (also called Norbonanol)
4.	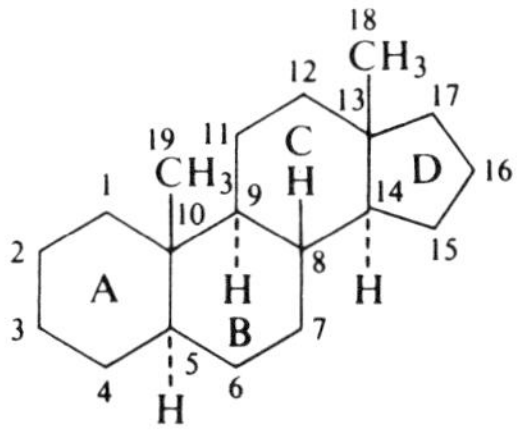 (structure)	Androstane[a]
5.	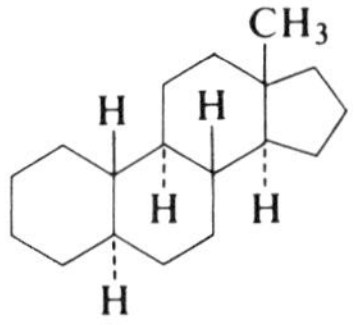 (structure)	19-Norandrostane
6.	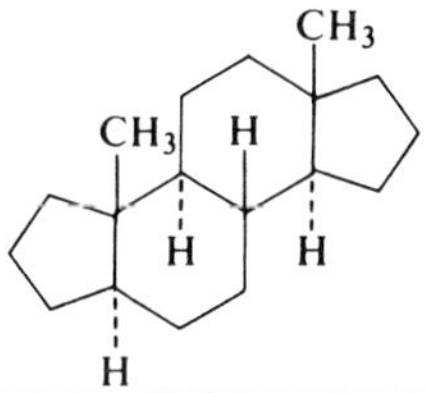 (structure)	A-Norandrostane

[a] These systems are numbered in a special way.

Engineering News article[8] should be consulted for nomenclature rules for other phosphorus containing compounds.

1.7 NOMENCLATURE FOR CIS-TRANS OR GEOMETRIC ISOMERS

The IUPAC has only recently published tentative rules for naming stereoisomers.[9] These rules are based largely on several practices that are firmly established and some of these are defined in the *Chemical Abstracts* article.[6] According to the IUPAC article,[9] *constitutional isomers* differ by the nature and sequence of bonding of their atoms, and *stereoisomers* differ only by the arrangement of their atoms in space. One type of

Table 1-23
CONJUNCTIVE NOMENCLATURE

1.	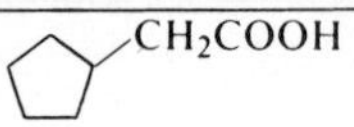 cyclopentane-CH$_2$COOH	Cyclopentaneacetic acid
2.	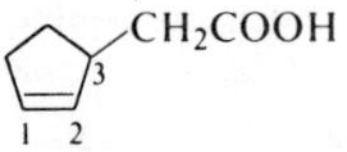 3-cyclopentene-CH$_2$COOH (positions 1,2,3)	3-Cyclopenteneacetic acid
3.	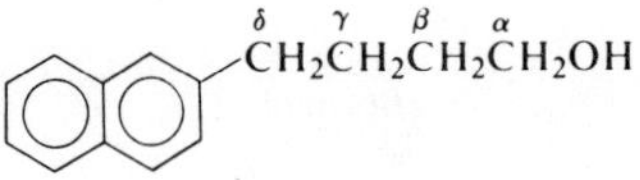 naphthalene-$\overset{\delta}{C}H_2\overset{\gamma}{C}H_2\overset{\beta}{C}H_2\overset{\alpha}{C}H_2OH$	2-Naphthalenebutanol
4.	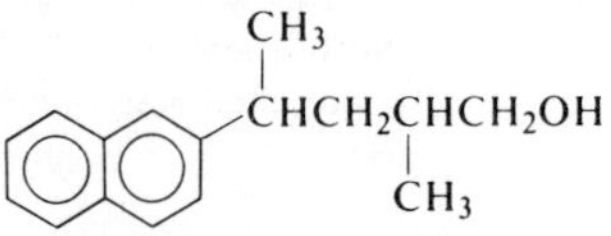 naphthalene-$\overset{\underset{\mid}{CH_3}}{C}HCH_2\overset{\underset{\mid}{CH_3}}{C}HCH_2OH$	β,δ-Dimethyl-2-naphthalene-butanol
5.	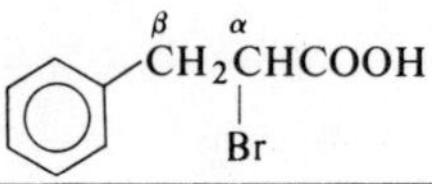 benzene-$\overset{\beta}{C}H_2\overset{\alpha}{C}HCOOH$ with Br	α-Bromobenzenepropionic acid

Table 1-24
REPLACEMENT NOMENCLATURE

CH$_3$OCH$_2$CH$_2$OCH$_2$CH$_2$OCH$_3$	2,5,8-Trioxanonane
CH$_3$OCH$_2$CH$_2$$\overset{\underset{\mid}{H}}{N}CH_2CH_2SCH_3$ 1 2 3 4 5 6 7 8 9	2-Oxa-8-thia-5-azanonane

Table 1-25
NOMENCLATURE FOR PHOSPHORUS COMPOUNDS

1.	CH$_3$CH$_2$$\overset{\overset{O}{||}}{P}(OH)_2$	Ethylphosphonic acid
2.	CH$_3$CH$_2$P(OH)$_2$	Ethylphosphonous acid
3.	Ph$_3$P	Triphenylphosphine
4.	Ph$_3$PO	Triphenylphosphine oxide
5.	(PhO)$_3$PO	Triphenyl phosphate
6.	(MeO)$_3$P	Trimethyl phosphite

stereoisomers is composed of *cis-trans* isomers, which differ in the position of atoms relative to a certain plane. These isomers are commonly called *geometric isomers*, and they usually result from restricted rotation about some bond. The most frequently encountered pairs of *cis-trans* isomers are those that result from restricted rotation about the carbon-carbon double bond. When two similar groups are on the same side of the double bond, the compound is called the *cis* isomer, and when they are on the opposite

1.7 Nomenclature for *Cis-Trans* or Geometric Isomers

side of the double bond, it is called the *trans* isomer. The terms *syn* and *anti* are used for specifying the geometry about certain double bonds such as the carbon-nitrogen double bond, and the terms *exo* and *endo* are frequently used to describe the geometric isomers in bicyclic systems. Examples of the use of these terms are given in Table 1-26.

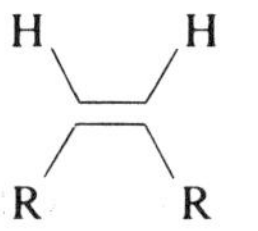

cis 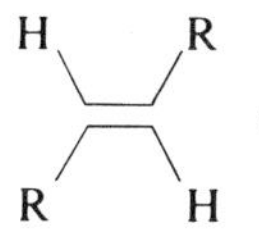   *trans*

Table 1-26
PREFIXES FOR *CIS-TRANS* OR GEOMETRIC ISOMERS

1.	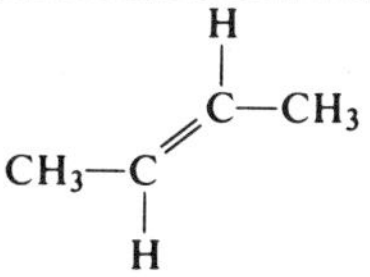 [CH₃—C(H)=C(H)—CH₃ structure]	*trans*-2-Butene; (*E*)-2-butene
2.	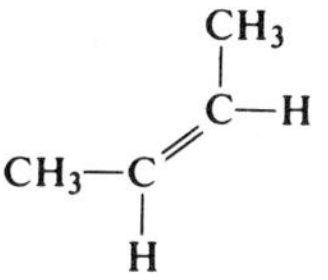 [CH₃—C(H)=C(CH₃)—H structure]	*cis*-2-Butene; (*Z*)-2-butene
3.	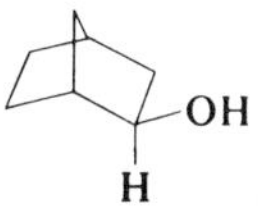 [norbornane with OH, H]	*exo*-2-Norbornanol
4.	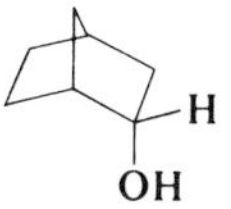 [norbornane with H, OH]	*endo*-2-Norbornanol
5.	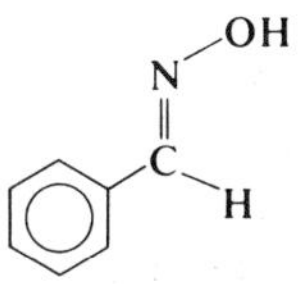 [PhCH=N-OH, H and OH on opposite]	*syn*-Benzaldehyde oxime; (*E*)-benzaldehyde oxime
6.	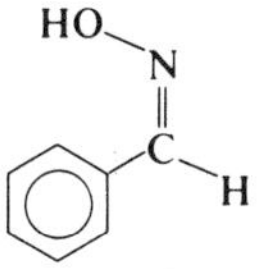 [PhCH=N-OH, H and OH on same side]	*anti*-Benzaldehyde oxime; (*Z*)-benzaldehyde oxime
7.	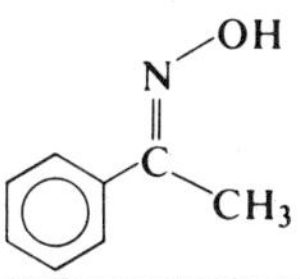 [PhC(CH₃)=N-OH]	Methyl *anti*-phenyl ketone oxime; (*E*)-methyl phenyl ketone oxime

In many cases, the specification of stereochemistry about a double bond becomes ambiguous with the use of the terms *cis-trans* or *syn-anti*. For example, does "*cis*-3-ethyl-3-octene" mean that the octene is *cis* or

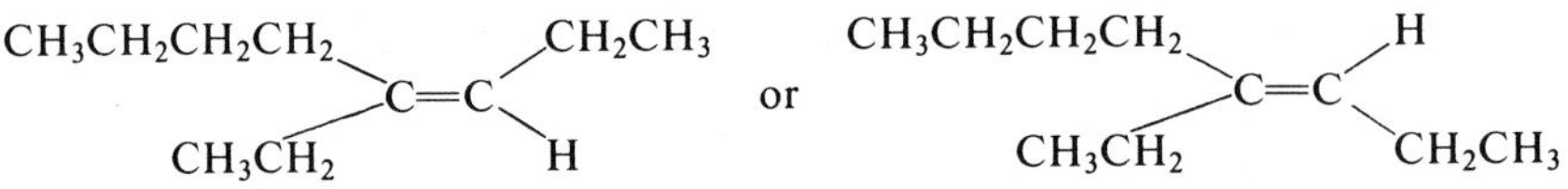

$$\begin{array}{c} CH_3CH_2CH_2CH_2 \\ CH_3CH_2 \end{array} C=C \begin{array}{c} CH_2CH_3 \\ H \end{array} \quad \text{or} \quad \begin{array}{c} CH_3CH_2CH_2CH_2 \\ CH_3CH_2 \end{array} C=C \begin{array}{c} H \\ CH_2CH_3 \end{array}$$

that the two equivalent ethyl groups on the double bond are *cis* to each other? In order to overcome this ambiguity, some *Chemical Abstract* workers[10] have proposed a set of rules for the use of two new descriptors, E and Z. The basis of this system is the assignment of higher priority to one of the two groups on each end of the double bond using the sequence rules of Cahn, Ingold, and Prelog.[11,12] These sequence rules[9,12] are used to assign priorities to groups in order to specify the absolute configuration of chiral molecules, i.e., ones that have a nonsuperimposable mirror image. This system is discussed in the next section. Briefly, the most important sequence rules may be summarized as follows:

Groups are arranged in order of decreasing atomic number of the atom by which they are bound. If two groups are bound by atoms having the same atomic number, then the groups should be ordered by comparison of the second set of atoms. If this fails, the third set should be used, etc. For double and triple bonds, the atom at the more remote end is counted twice or thrice.

Thus, the following groups are arranged in decreasing priority.

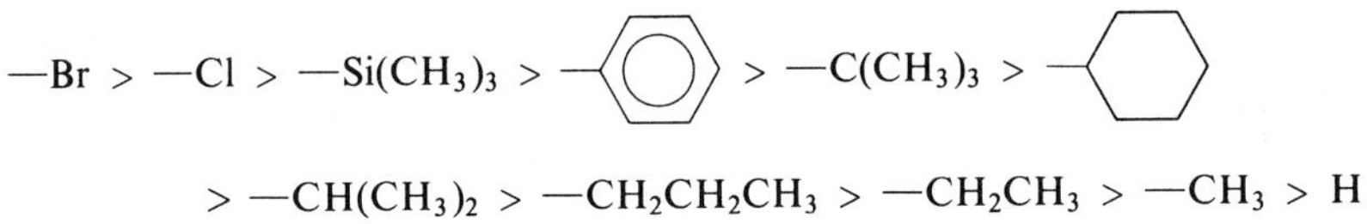

$$-Br > -Cl > -Si(CH_3)_3 > -C_6H_5 > -C(CH_3)_3 > -C_6H_{11}$$
$$> -CH(CH_3)_2 > -CH_2CH_2CH_3 > -CH_2CH_3 > -CH_3 > H$$

There are many other rules which give the priorities of stereoisomeric groups, etc., but these will not be discussed here. Tables which give the priorities of many groups can be found in the IUPAC article.[9]

If the two groups of higher priority are on the same side of the double bond, then the isomer is assigned a Z, which is derived from the German word for together, *zusammen*. On the other hand, if the two groups of higher priority are on opposite sides of the double bond, then the isomer is assigned an E, which comes from the German word *entgegen*, meaning opposite. In Table 1-26 each double bond isomer is assigned an E or Z. From these examples and the case of the 3-ethyl-3-octene isomers mentioned above, the superiority of the use of E and Z is quite apparent. The aldehyde oxime isomers especially illustrate the superiority of the E-Z descriptors since only by definition does one know whether *syn* or *anti* is used with respect to the hydrogen or the hydrocarbon group (the accepted practice is, in fact, to use *syn* or *anti* with respect to the hydrogen atom for aldehyde oximes).

Although the extensive use of *cis*, *trans*, *syn*, and *anti* in the past and present literature ensures their use in the future, the superiority of the *E-Z* descriptors will probably lead to their general acceptance and extensive use in the future.

1.8 NOMENCLATURE FOR SPECIFICATION OF ABSOLUTE CONFIGURATION OF ASYMMETRIC CARBON ATOMS

Often it is desirable to be able to specify the absolute configuration of an asymmetric carbon atom. A system to do this, which is well-accepted and in general use, has been developed by Cahn, Ingold, and Prelog. Cahn, Ingold, and Prelog have given rules for specifying the absolute configuration of many types of chiral molecules.[9,11] Only the system for specifying the absolute configuration of an asymmetric carbon atom will be given here.

The four groups attached to the asymmetric carbon atom are assigned priorities using the sequence rules that were presented in the last section. For example, let the four groups attached to the asymmetric carbon atom be called W, X, Y, and Z and let them decrease in priority in that order. One then sights along the carbon-Z bond where Z is the group of lowest

Table 1-27

R- AND S- PREFIXES FOR MOLECULES THAT CONTAIN ASYMMETRIC CARBON ATOMS

Structure	Name
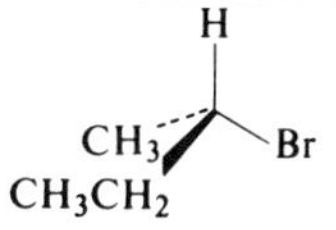 CH₃, CH₃CH₂, H, Br on C	(*S*)-2-Bromobutane
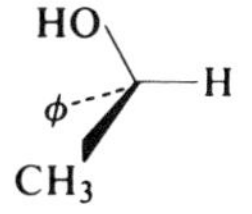 HO, φ, H, CH₃ on C	(*S*)-1-Phenylethanol
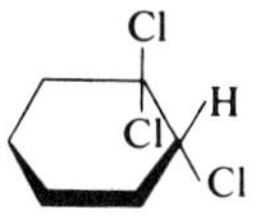 Cyclohexane with Cl, Cl, H, Cl	(*R*)-1,1,2-Trichlorocyclohexane
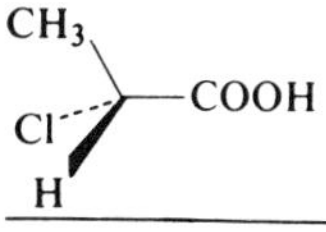 CH₃, Cl, H, COOH on C	(*R*)-α-Chloropropionic acid

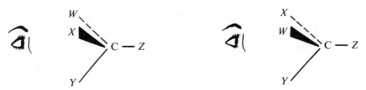

R (rectus, right or clockwise)   S (sinister, left or counterclockwise)

priority. If the groups W, X, and Y are arranged in a clockwise sequence, then the asymmetric carbon atom has the *R* configuration. Conversely, the counterclockwise arrangement of W, X, and Y is assigned the *S* configuration. These descriptors, *R* and *S*, come from the Latin words for right and left, *rectus* and *sinister*, respectively. Often one of the groups attached to the asymmetric carbon atom is a hydrogen atom and in these cases, it will, of course, be the group of lowest priority. The examples in Table 1-27 illustrate these rules.

PROBLEMS

1. Draw all the constitutional isomers and stereoisomers of octane, C_8H_{18}, whose longest chain of carbon atoms contains six atoms. Indicate the pairs of isomers which are enantiomers.

2. Draw all the constitutional isomers and stereoisomers with the formula $C_3H_3Cl_3$ that contain a cyclopropane ring.

3. Draw all the constitutional isomers with the formula C_4H_8O that contain a three-membered ring.

4. Draw all the structures of reasonable compounds having the formula C_4H_7NO that possess a four-membered ring and a carbonyl group.

5. Draw all the different dibromonaphthalenes.

6. Name the following hydrocarbons.

(a) CH$_3$CH$_2$CH$_2$CHCH$_2$CH$_2$CH$_2$CH$_3$
       |
       CH$_2$CH$_3$

(b) CH$_3$CH$_2$CH$_2$CHCH$_2$CH$_2$CH$_2$CH$_3$
       |
       CH=CH$_2$

(c) CH$_3$CHCH$_2$—CH—CH$_2$—CH—CH$_2$—CH—CH$_2$CH$_2$CH$_3$
       | | | |
       CH$_2$CH$_3$ CH(CH$_3$)$_2$ C(CH$_3$)$_3$ CH(CH$_3$)$_2$

(d) CH$_3$CH=CH—C—CH—CH=CH$_2$
               | |
               CH$_2$CH$_2$CH$_2$CH$_3$ (above C)
               CH$_3$ CH$_3$

Problems

(e) [structure: benzene with two adjacent CH₃ groups]

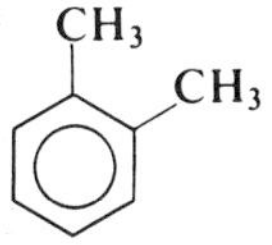

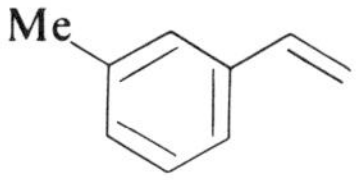

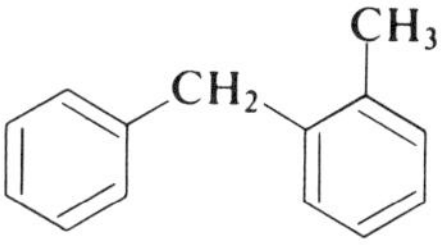

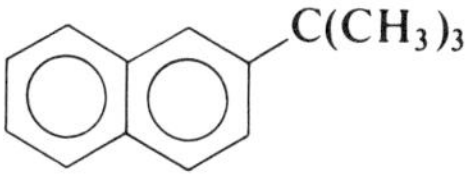

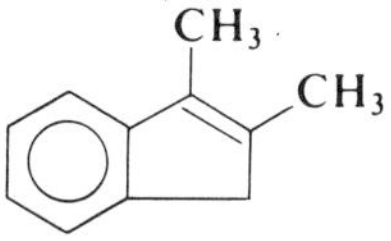

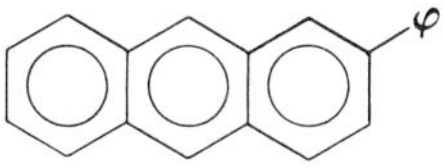

(f) [structure: cyclopentyl-CH₂CH=CH₂]

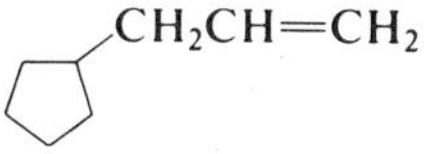

(g) [structure: phenyl-CH₂CH₂CH₂-cyclohexyl]

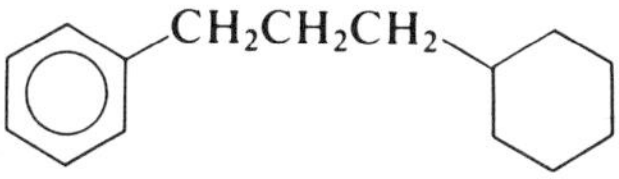

(h) Me-[structure: m-methyl styrene]

(i) [structure: phenyl-CH₂-(2-methylphenyl)]

(j) [structure: naphthalene with C(CH₃)₃ substituent]

(k) [structure: indene with two CH₃ groups]

(l) [structure: anthracene with φ substituent]

(m) [structure: norbornane with φ substituent]

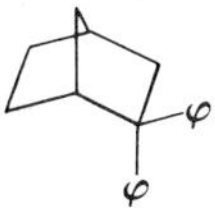

(n) [structure: bicyclic alkene with two CH₃ groups on bridge]

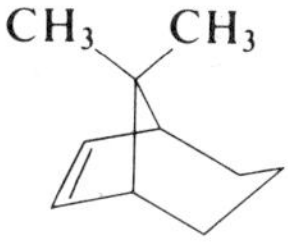

(o) [structure: bicyclic structure with two CH₃ groups]

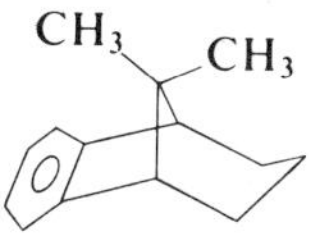

(p) φ–C(CH₃)₂–C(CH₃)₂–φ

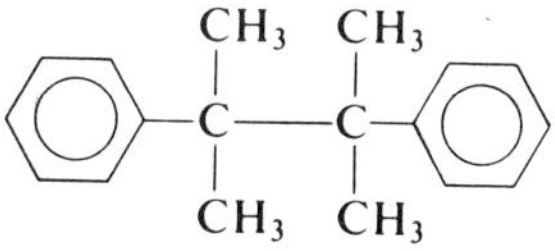

(q) [structure: cyclohexyl-cyclobutyl]

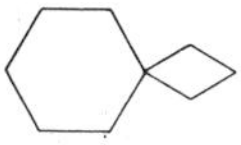

7. Give two IUPAC names (i.e., use different nomenclature systems) for the following systems).

(a) [phenyl-Br]

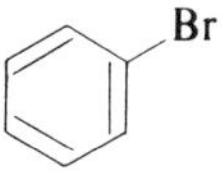

(b) [phenyl-CH₂COCH₂CH₃]

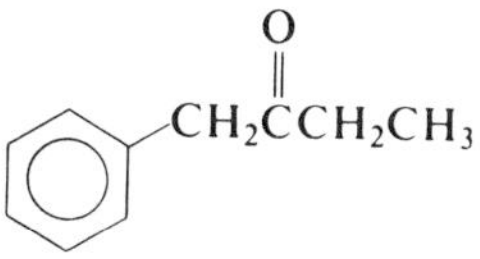

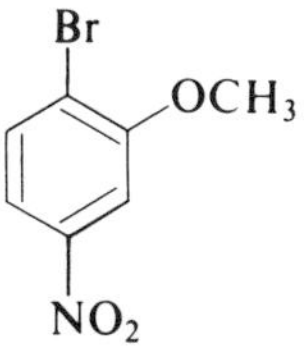

(c) [naphthyl-CH₂CH=O]

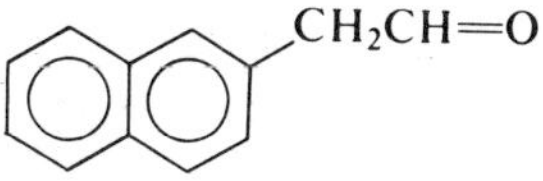

(d) ClCH₂CH₂CH₂COOH

(e) [cyclopropyl]–CH₂CN

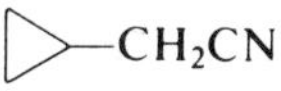

8. Name the following compounds.

(a) [phenyl-CH(CH₃)CH₂CH₂OH]

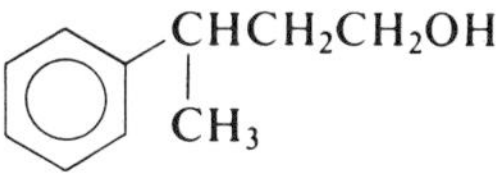

(b) [2-bromo-3-methoxy-5-nitrobenzene: Br, OCH₃, NO₂ substituted benzene]

(c) OH O
   CH₃CHCH₂CH₂CCH₃

(d) CH₃SO₂—φ

(e) CH₃SO—φ

(f) φ₂NH

(g) CH₃CH₂CH₂CH₂CH₂N₃

(h) 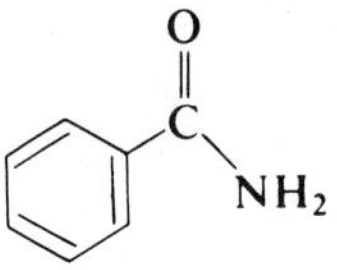 [structure: benzamide, PhC(=O)NH₂]

(i) 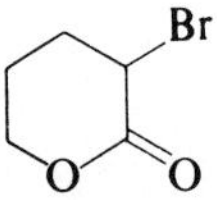 [structure: 3-bromo-tetrahydro-2H-pyran-2-one]

(j)
[structure: 1,3-dimethyl tropylium cation with Br⁻]

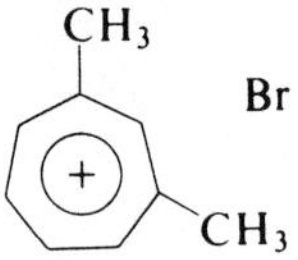

(k) CH₃CH₂CH₂CH₂Li

(l) 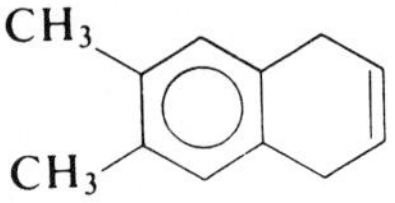 [structure: 2,3-dimethyl-1,2-dihydronaphthalene-like]

(m) 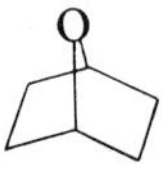 [structure: bicyclic ketone, norcamphor-like]

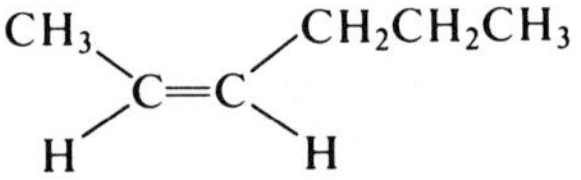

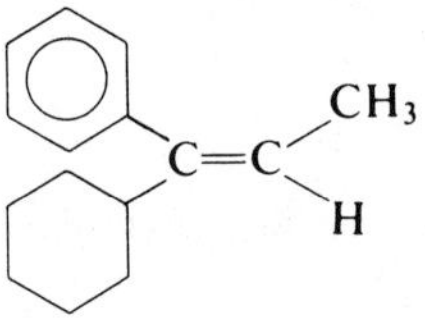

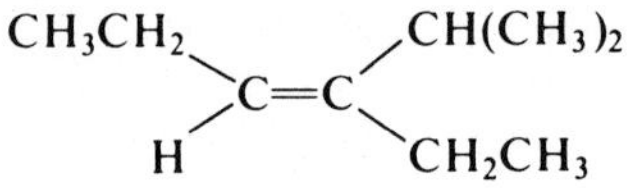

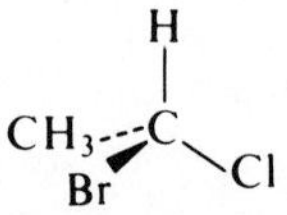

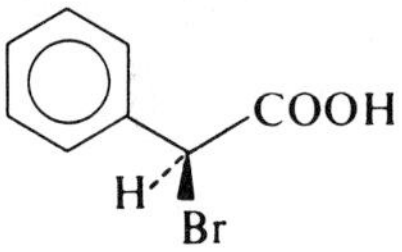

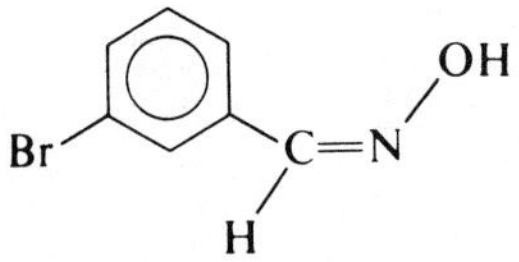

9. Give a prefix that denotes the stereochemistry of the following compounds.

(a) 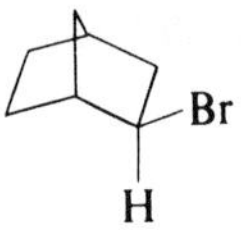 [norbornyl-Br with H]

(b) CH₃\ /CH₂CH₂CH₃
    C=C
    H / \ H

(c) [Ph and cyclohexyl on C=C with CH₃ and H]

(d) CH₃CH₂\ /CH(CH₃)₂
    C=C
    H /   \CH₂CH₃

(e) H
    |
    CH₃--C
    Br \Cl

(f) [Ph–CH(Br)–COOH with H wedge]

(g) [3-bromophenyl–CH=N–OH]

(h) [β-lactam-like ring with C=N–CH₃]

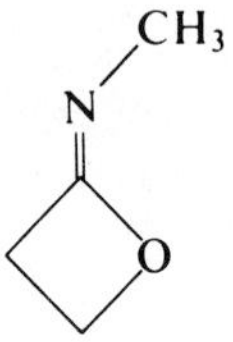

10. Give structures for the following compounds.
(a) Isopropylidenecyclopentane
(b) Phenanthrene
(c) Bicyclopentyl
(d) 2,2-Dimethylspiro[3.5]nonane
(e) 3-Chlorohexanal
(f) Cyclopentyl phenyl ketone
(g) 9-Bicyclo[3.2.1]octanecarboxylic acid
(h) cis-3-Nonene oxide
(i) 2-Azabicyclo[2.2.2]octane

Problems

(j) Fluorene
(k) Propylidenecyclobutane

11. Name the following compounds.

(a) CH₃COCH₂—CH(—CH₂CH₂—CH₂COCH₃)—CH₂CH₂CH₂—CH(OH)—CH₂CH₃

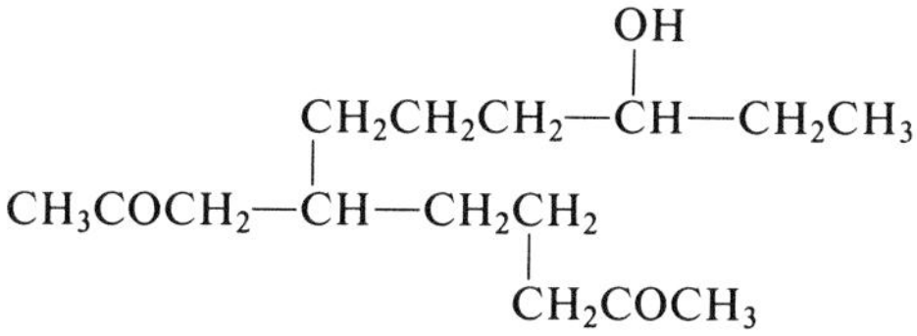

(b) C₆H₅—P(CH₂CH₃)(CH₃)

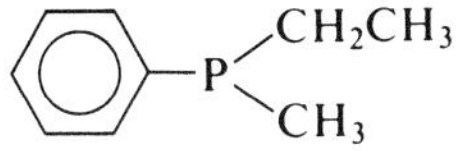

(c) bicyclic structure with COOH

(d) norbornene structure

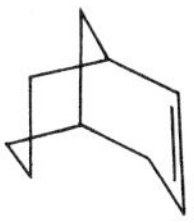

(e) 1,1-dimethylcyclohexyl cation (CH₃, CH₃, +)

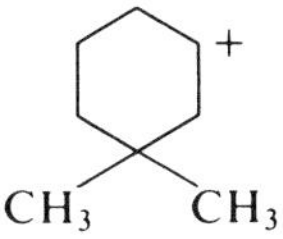

(f) 3-chlorophenyl-CH₂COOH

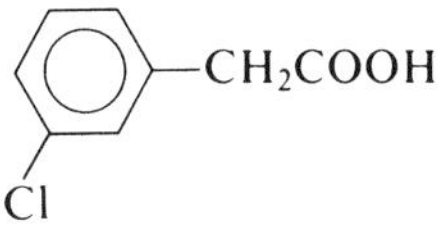

12. Name the following compounds, being sure to specify stereochemistry.

(a) cyclic diene structure

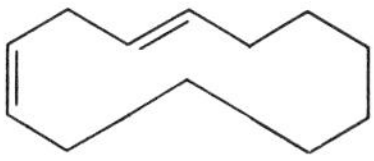

(b) (CH₃CH₂)(CH₃)C=C(CH₃)(H)

(c) Ph—C(=N—OH)—C₆H₄Cl

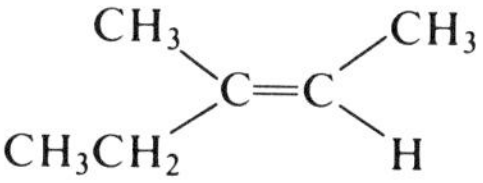

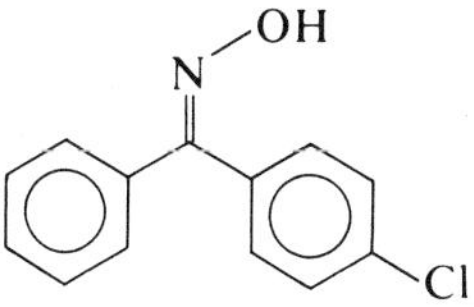

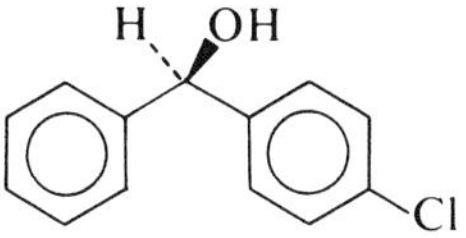

(d) Ph—CH(OH)—C₆H₄Cl (H, OH wedge)

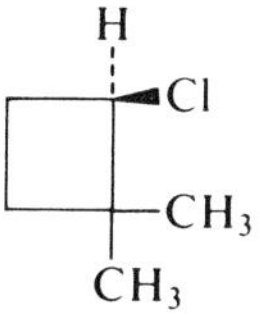

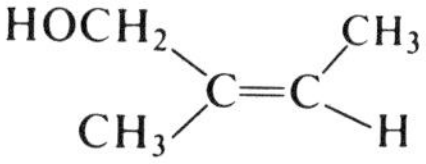

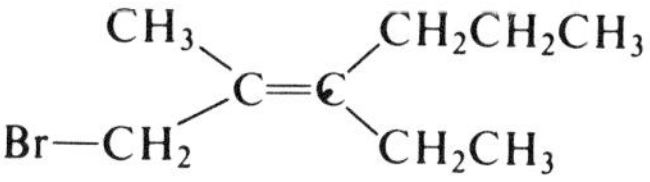

(e) cyclobutane with H, Cl, CH₃, CH₃ substituents

(f) (HOCH₂)(CH₃)C=C(CH₃)(H)

(g) (CH₃)(BrCH₂)C=C(CH₂CH₂CH₃)(CH₂CH₃)

(h) cyclopentane with OH, H, Cl, Cl

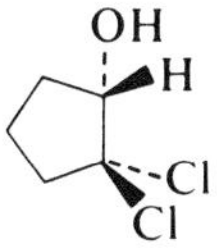

13. Give two IUPAC names (i.e., use different nomenclature systems) for the following compounds.

(a) CH₃CH₂C(=O)CH₂—C₆H₄—NO₂

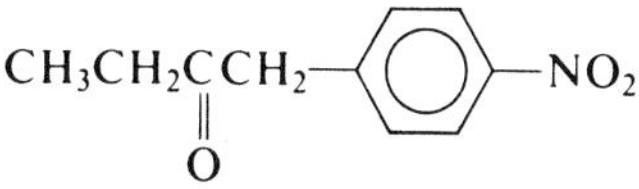

(b) cyclopentyl—OH

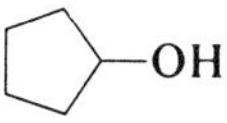

(c) cyclopentene oxide

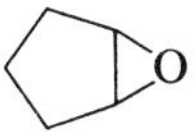

(d) CH₃CH₂OCH₂CH₂CH₂CH₂CH₃

14. Give the name of the nomenclature systems (i.e., substitutive, etc.) that were used to generate the following names.

(a) Cyclopropanepropanol

(b) 4-Azaspiro[2.4]heptane

(c) D-Norcholesterol

(d) 9,10-Dihydroanthracene

REFERENCES

1. *J. Amer. Chem. Soc.*, **82,** 5545 (1960).

2. *Pure and Applied Chemistry*, **11,** 1 (1965).

3. *Nomenclature of Organic Chemistry Sections A and B*. London: Butterworth & Co., 1957.

4. *Nomenclature of Organic Chemistry Section C*. London: Butterworth & Co., 1965.

5. *J. Amer. Chem. Soc.*, **55,** 3905 (1933).

6. *Chemical Abstracts*, Subject Index, Vol. **56,** 1962.

7. D. Eckroth, *J. Org. Chem.*, **32,** 3362 (1967).

8. *Chem. Eng. News*, **30,** 4515 (1952).

9. *J. Org. Chem.*, **35,** 2849 (1970).

10. J. E. Blackwood, C. L. Gladys, K. L. Loening, A. E. Patrarca, J. E. Rush, *J. Amer. Chem. Soc.*, **90,** 509 (1968).

11. R. S. Cahn, C. K. Ingold, and V. Prelog, *Angew. Chem., Intern. Ed. Engl.*, **5,** 385 (1966).

12. _____, *Experientia*, **12,** 81 (1956).

2
The Relationship Between Physical Properties and Molecular Structure

2.1 INTRODUCTION

In this chapter, we will be concerned with the ways in which certain structural features of molecules affect their physical properties. First, we should have a clear idea about what a physical property is. A *physical property* of a substance is a macroscopic characteristic interaction of that substance with its environment in a physical sense as opposed to a chemical sense; i.e., the interaction does not irreversibly alter the structure of the substance. The physical properties that we will be concerned with are boiling points, melting points, dipole moments, solubility, chromatographic behavior, and infrared, nuclear magnetic resonance, and electronic spectra. Our goal will be to interpret these macroscopic physical properties in terms of the molecular structure.

Two other extremely important types of spectrometry, mass spectrometry and electron spin resonance, will not be discussed in this chapter since neither type involves a physical property of a molecule in the sense defined above. In mass spectrometry, stable molecules are converted to relatively unstable ions. In electron spin resonance, only paramagnetic compounds give spectra and most paramagnetic organic compounds are relatively unstable species derived from stable organic molecules. Mass spectrometry and electron spin resonance are discussed in G. A. Sim, K. L. Rinehart, Jr., R. O. C. Norman, and B. C. Gilbert's book in this series, *X-Ray Crystallography, Mass Spectrometry, and Electron Spin Resonance of Organic Compounds*.

2.2 BOILING POINTS

Most pure liquid substances boil at a characteristic temperature for a given pressure. The macroscopic phenomenon that occurs at a liquid's boiling point is that it is rapidly turning from a liquid to a gas. On a molecular level, the molecules in the liquid phase have an escaping tendency that results in a vapor pressure that equals the applied pressure. This simple microscopic picture of a liquid at its boiling point enables us to make some predictions. For example, if we raise the applied pressure, the boiling point of any (well-behaved) liquid should rise. Also a liquid at a temperature just below its boiling point should tend to disappear or evaporate quicker than it would at a lower temperature, since the escaping tendency of the molecules would be higher at a higher temperature even though that temperature is not sufficient to generate a vapor pressure equal to the applied pressure. Thus, this simple microscopic picture enables us to readily understand macroscopic phenomena such as the facts that water boils at a higher temperature in a pressure cooker and lighter fluid that is spilled on the floor disappears quicker than water that is spilled on the floor. We turn now to the question of how we can predict the escaping tendency of a liquid based on the structure of the molecules that make up the liquid.

Functional groups can have an enormous effect on boiling points, as we shall see, but before we consider the effect of functional groups, let us consider molecules that are devoid of functional groups in order to see how the general structure of a molecule affects its boiling point.

Four important molecular features that influence the escaping tendency of a molecule are its weight, size, shape, and flexibility. In most cases, an increase in weight leads to an increase in size and usually these two effects are difficult to separate. Thus at one atmosphere pressure, pentane (C_5H_{12}) boils at 36°, hexane (C_6H_{14}) boils at 69°, and heptane (C_7H_{16}) boils at 98°. The normal alkanes all behave as expected and each additional —CH_2— group causes a 20–30 degree increase in boiling point as a result of additional size and weight. But let us consider an alkane of different shape, since the normal alkanes are all chain-like molecules with a high degree of flexibility. What about a compound like neopentane, $(CH_3)_4C$? From molecular weight considerations alone, one would predict that neopentane should boil at about 35°, like its isomer *n*-pentane. Moreover, the equal densities of the two liquids suggest equal size. However, neopentane boils at 10°! Actually, this lower boiling point is quite reasonable, since the rigid, sphere-like neopentane molecules should be able to escape the condensed phase easier than the chain-like *n*-pentane molecules. The intermolecular attractions which prevent molecules from escaping the surface of a liquid increase as the area of contact between the molecules increases. For this same reason, it is easier to keep a shovel full of grass-clippings together than a shovel full of leaves, even though a leaf is much heavier and larger than a grass-clipping. Applying this picture, one predicts that more rigid, sphere-like molecules will boil lower than their more flexible isomers and, indeed, series of isomers such as

2.2 Boiling Points

2,2,4-trimethylpentane (bp 99°), 2-methylheptane (bp 116°), and octane (bp 126°) substantiate this generalization.

Since weight and size generally increase together, there is not much need to separate their effects; but comparison of neopentane [$(CH_3)_4C$, bp 10°, MW 72] with tetramethylsilane [$(CH_3)_4Si$, bp 27°, MW 88] shows that increase of molecular weight alone leads to an increased boiling point and comparison of 3-ethylhexane [$(CH_3CH_2)_2CHCH_2CH_2CH_3$, bp 119°, MW 114] with triethylsilane [$(CH_3CH_2)_3SiH$, bp 107°, MW 116] suggests that an increase of size alone also leads to an increased boiling point.

Now let us consider molecules with functional groups. If the functional group has no special interactions with itself or the rest of the molecule, then the molecular weight, size, shape and flexibility should be the controlling features of boiling points. Indeed pairs of compounds such as diethyl ether (bp 35°, MW 74) and pentane (bp 36°, MW 72) or butyl ethyl ether (bp 91°, MW 102) and heptane (bp 98°, MW 100) support this idea. But what about ethyl methyl ketone, C_4H_8O, MW 72, which boils at 80° instead of around 35° as do diethyl ether and pentane? Let us consider the structure of the carbonyl group. The three groups around the carbonyl carbon are arranged approximately at the apices of an equilateral triangle with the carbonyl carbon in the center of the triangle. The electronegative

$$\begin{array}{c} R \\ \diagdown \\ C=O \\ \diagup \\ R \end{array}$$

oxygen makes the dipolar resonance form of the carbonyl important.

$$\begin{array}{c} R \\ \diagdown \\ \overset{+}{C}-\overset{-}{O} \\ \diagup \\ R \end{array}$$

This can be represented in the language of resonance by enclosing the two forms in brackets and separating them with a double-headed arrow, or by

$$\left[\begin{array}{c} R \\ \diagdown \\ C=O \\ \diagup \\ R \end{array} \longleftrightarrow \begin{array}{c} R \\ \diagdown \\ \overset{+}{C}-\overset{-}{O} \\ \diagup \\ R \end{array} \right]$$

using δ's to represent charge separations. The important point is that the

$$\begin{array}{c} R \\ \diagdown \delta^+ \delta^- \\ C=O \\ \diagup \\ R \end{array}$$

carbonyl group does have unsymmetrical charge distribution which leads to a permanent dipole. The dipole is represented by an arrow with a

$$\begin{matrix} R \\ \diagdown \\ C{=}O \\ \diagup \\ R \end{matrix} \;\longmapsto$$

crossed tail pointing in the direction of the negative charge. In a condensed phase, molecules with dipoles will tend to attract each other since their positive portions will attract negative portions of other molecules and vice versa. The dipoles of a small section of liquid might look as follows:

$$\begin{matrix} \longmapsto & \longmapsto \\ \longmapsto & \longmapsto \\ \longleftarrowtail & \longleftarrowtail \\ \nearrow\!\!\!\!\!\mid & \nwarrow\!\!\!\!\!\mid \end{matrix}$$

The dipole of the carbonyl group, then, easily explains the high boiling point of ethyl methyl ketone.

So, in general, compounds that contain functional groups that give rise to a dipole should boil higher than compounds without dipoles (even if the molecule contains two polar groups with opposing dipoles that cancel, since each dipole is still important on a molecular level). All of the functional groups that contain a carbonyl group have permanent dipoles.

Another very polar functional group is the nitro group. The collection

$$\left[R-\overset{+}{N}\!\!\begin{matrix}\diagup O \\ \diagdown O_-\end{matrix} \quad \longleftrightarrow \quad R-\overset{+}{N}\!\!\begin{matrix}\diagup O^- \\ \diagdown O\end{matrix} \right]$$

of three electronegative atoms leads to a strong permanent dipole which

$$R-NO_2 \\ \longmapsto$$

readily accounts for the high boiling point of nitroethane (bp 115°, MW 75) compared to the structurally similar 2-methylbutane (bp 28°, MW 72) or even to 2-methylpropanal (bp 62°, MW 72).

Another very important characteristic of molecules, the ability to hydrogen bond, is illustrated by butanol, which boils at 118° compared to its isomer diethyl ether which boils at 35°. The hydrogen bond is a relatively strong (*ca.* 5 kcal/bond) intermolecular interaction that accounts for the high boiling points of compounds that contain a relatively acidic hydrogen and a basic site. Thus alcohols, amines, carboxylic acids, and

$$B\text{---}H\text{---}A$$

amides all have relatively high boiling points due to hydrogen bonding.

2.2 Boiling Points

Table 2-1

ILLUSTRATIONS OF THE EFFECT OF HYDROGEN-BONDING FUNCTIONAL GROUPS ON BOILING POINTS

C_2 derivative	bp,°	MW	Comparable alkane	bp,°	MW
CH_3CH_2OH	79	46	$CH_3CH_2CH_3$	−45	44
$CH_3CH_2NH_2$	17	45	$CH_3CH_2CH_3$	−45	44
CH_3COOH	119	60	$CH_3\overset{\underset{\mid}{CH_3}}{C}HCH_3$	−1	58
CH_3CONH_2	221	59	$CH_3\overset{\underset{\mid}{CH_3}}{C}HCH_3$	−1	58

Table 2-1 illustrates this by giving the boiling point of the C_2 derivative and that of an alkane of similar structure and molecular weight.

Physical properties are not only useful in characterizing compounds, they are also useful in helping to determine the structure of compounds. For example, if a chemist has two liquids which boil at 126° and 116°, and he knows that one is octane and the other is isoöctane, he will guess correctly that the lower boiling liquid is the branched alkane without looking up their boiling points. Of course, even before modern spectral methods, there were chemical ways to determine the structures of octane and isoöctane unequivocally, but a much more difficult type of isomerism to deal with is *cis-trans* (geometrical) isomerism. How would one determine which $C_2H_2Cl_2$ isomer is *cis* and which one is *trans*? Looking at their formulas, one concludes that the *cis* isomer should possess a perma-

zero dipole moment

nent dipole moment since the two electronegative chlorine atoms are pulling electrons in the same direction. This would support the assignment of the *cis* structure to the isomer that boils at 60° and the *trans* structure to the isomer that boils at 48°. Indeed very strong support of this assignment is given by the fact that a charged hair comb will attract a stream of the higher boiling liquid but not affect a stream of the lower boiling liquid. This latter experiment, coupled with an instrumental determination of a permanent dipole moment in the higher boiling isomer and none in the lower boiling isomer, establishes the *cis-trans* assignment unequivocally. Nevertheless, as we discuss other physical properties, we will come back to this pair of compounds and gain further support for their structures.

2.3 MELTING POINTS

Another important physical property of a compound is its melting point, the temperature at which the liquid and solid exist in equilibrium. For most crystalline compounds, the melting point is quite sharp and serves as a means of characterizing the substance. Most organic compounds that are solids melt below 200°, and many simple apparatus are available for taking melting points. Usually a small sample of the compound is placed in a capillary tube with one sealed end. The tube is then placed in a liquid bath which is slowly and evenly heated. When the compound melts, the temperature of the bath corresponds to the melting point.

In this section, we are going to be concerned with correlating molecular structure with melting points, much as we were concerned with correlating molecular structure with boiling points. A glance at melting points for the four C_{10} hydrocarbons given in Table 2-2 immediately shows that molecular weight is of little or no concern with respect to melting points. Indeed, consideration of the microscopic processes which are going on while a crystal is melting leads to the conclusion that size, flexibility, and shape are much more important than weight. In a crystal, the molecules are all packed in some orderly fashion. Of course, the molecules are close to one another and are usually not tumbling or undergoing any translational movements. In the liquid phase, the molecules are about as close to each other as in the solid phase but instead of being oriented, they are tumbling and moving past one another. Thus, molecular weight should be an unimportant factor when a solid melts, because the molecules are not escaping a condensed phase to enter a void but are escaping an oriented, condensed phase to enter a disoriented, condensed phase. Quite naturally, molecular size, flexibility, and shape should be the important factors in determining the melting point of a substance. Also, other intermolecular forces, such as dipole attractions and hydrogen

Table 2-2

MELTING POINTS OF FOUR C_{10} HYDROCARBONS

Name	Structure	Formula	MW	Melting point,°
Decane	$CH_3(CH_2)_8CH_3$	$C_{10}H_{22}$	142	−29.7
Durene		$C_{10}H_{14}$	134	79
Camphane		$C_{10}H_{18}$	138	158
Adamantane		$C_{10}H_{16}$	136	268

2.3 Melting Points

bonding, which should lead to preferred orientations should and do affect melting points.

In thinking about the effect of molecular structure on melting points, consideration of the solid phase is much more important than consideration of the liquid phase, since the crystal energy or packing energy will be the biggest variable from substance to substance. Thus, the molten state of decane and durene will be comparable in stability but the solid state of durene will be much more stable than that of decane since the rigid flat molecules can be more easily packed. The more difficult it is to go from randomly oriented molecules to highly ordered packed molecules, the less stable will the crystal be.

In the discussion that follows, we will try to come up with a most reasonable structure for a given crystal. Actually, the structures we arrive at will not just be theoretical, since modern X-ray crystallography has found that indeed most molecules pack in these "most reasonable" structures.[1]

In order to avoid complications due to dipolar intermolecular attractions and intermolecular hydrogen bonding, hydrocarbons will be considered to illustrate the effects of size, flexibility, and shape on melting points. In Table 2-3 are presented the melting points of various normal alkanes. In general, as the alkane chain increases in length, the melting point increases, but an increase in length does not always lead to an increase in melting point. In going from methane to ethane, the molecular weight and length are both doubled, but the melting point drops! Obviously the higher melting point of methane is a result of its higher sym-

Table 2-3

MELTING POINT OF NORMAL ALKANES

Alkane	Number of carbon atoms	mp,°
Methane	1	−182
Ethane	2	−183
Propane	3	−190
Butane	4	−138
Pentane	5	−130
Hexane	6	− 94
Heptane	7	− 91
Octane	8	− 57
Nonane	9	− 54
Decane	10	− 30
Eicosane	20	37
Heneicosane	21	40
Docosane	22	44
Tricosane	23	48
Tetracosane	24	51
Triacontane	30	66

metry. A ball-like molecule packs more readily than a sausage-like molecule, since the ball-like molecule can be oriented in many different ways and still fit. The propane molecule, which contains a bent carbon skeleton, packs with even more difficulty than ethane and consequently has a melting point even lower than that of ethane. The increased difficulty in packing methane, ethane, and propane molecules is easy to appreciate if one considers the increased care it would take to pack baseballs, sausages, and boomerangs in boxes.

After butane, the alkanes all have approximately the same zig-zag shape[1] and simply increase in length. From C_{20} to C_{30}, the alkanes are so long that only small regular increases in melting points per carbon are found.

The normal alkanes certainly show that size is important, but the four C_{10} hydrocarbons in Table 2-2 strongly suggest that flexibility and shape are more important. The melting points of the cycloalkanes, given in Table 2-4, illustrate the importance of flexibility. All of the cycloalkanes have a cyclic shape, but the different numbers of carbon atoms in the rings lead to a great difference in flexibility. Thus, the four-member ring is relatively rigid and has a higher melting point than cyclopentane, which is flexible.* The fact that cyclohexane melts 101° higher than cyclopentane is undoubtedly a result of the very rigid and stable chair form of cyclohexane. The increased flexibility of cycloheptane leads to a melting point drop, whereas the relatively rigid crown form of cyclooctane causes a high melting point.

Table 2-4
MELTING POINTS OF CYCLOALKANES

Alkane	mp°	Stable conformation	
□	−50	Planar or slightly puckered	
⬠	−94	Flexible	
⬡	7	(chair shape)	"chair"
⬣	−12	Flexible	
⯃	14	(crown shape)	"crown"

In general, the higher the symmetry of a rigid molecule, the higher its melting point. Pairs of compounds such as anthracene and phenanthrene illustrate this point.

* N. L. Allinger and J. Allinger, *Structure of Organic Molecules*, in this series, p. 104.

2.3 Melting Points

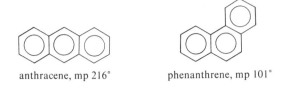

anthracene, mp 216°   phenanthrene, mp 101°

Just as in the case of boiling points, dipoles lead to strong intermolecular attractions. Thus, a molecule with a permanent dipole might be expected to form a crystal with the dipoles oriented as shown in Fig. 2.1. Obviously, these additional attractions will lead to a higher melting solid.

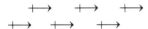

Fig. 2.1 Possible orientation of dipoles in a crystalline solid.

The strong dipole of cyclohexanone readily explains its high melting point compared to methylene cyclohexane.

cyclohexanone, mp −16°   methylenecyclohexane, mp −107°

Intermolecular association can occur even if the compound does not have an overall dipole moment but has two or more cancelling dipoles. The melting points of the *p*-dimethyl-, *p*-dichloro-, and *p*-dinitrobenzenes show that as the dipole of the carbon-substituent bond increases, the melting point increases, even though the resulting dipole moment of

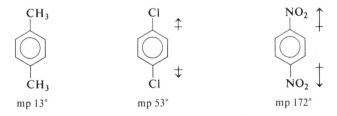

mp 13°   mp 53°   mp 172°

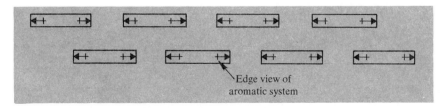

Fig. 2.2 Possible crystal structure of molecules that contain cancelling dipoles.

the molecule is zero. Packing such as illustrated in Fig. 2.2 has been shown to be important in these cases.

In Table 2-5 are presented melting points of compounds containing various functional groups and hydrocarbons of comparable size, shape, and weight. In general, the higher melting compounds possess functional groups with a greater dipole moment. The higher melting points of compounds that contain nitro, sulfoxide, or sulfone groups are consistent with the importance of dipolar resonance structures for these groups.

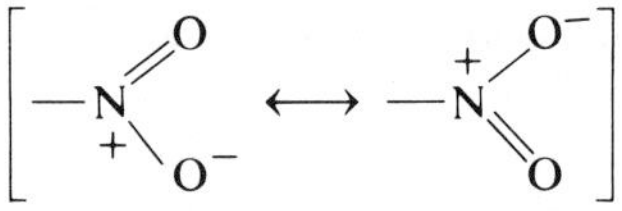

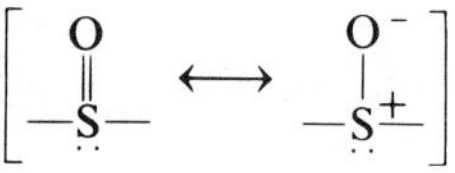

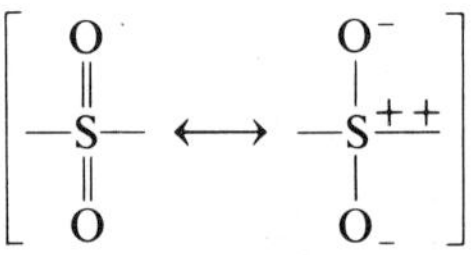

$$\left[-\overset{+}{N}\diagup_{\diagdown O^-}^{\diagup\!\!\!O} \longleftrightarrow -\overset{+}{N}\diagup_{\diagdown O}^{\diagup\!\!\!O^-} \right]$$

$$\left[-\overset{O}{\underset{..}{\overset{\|}{S}}}- \longleftrightarrow -\overset{O^-}{\underset{..}{\overset{|}{S^+}}}- \right]$$

$$\left[-\overset{O}{\underset{\overset{\|}{O}}{\overset{\|}{S}}}- \longleftrightarrow -\overset{O^-}{\underset{\overset{|}{O_-}}{\overset{|}{S^{++}}}}- \right]$$

Table 2-5
MELTING POINTS OF POLAR ORGANIC COMPOUNDS AND STRUCTURALLY SIMILAR HYDROCARBONS

Compound	mp, °	Compound	mp, °
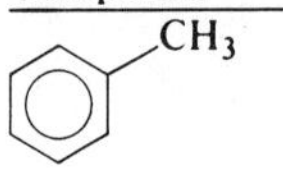 C₆H₅—CH₃	−95	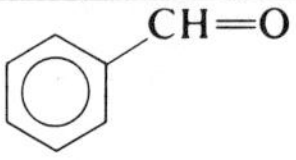 C₆H₅—CH=O	−56
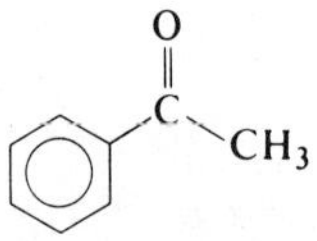 C₆H₅—C(=O)—CH₃	20	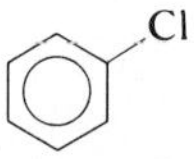 C₆H₅—Cl	−45
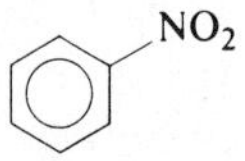 C₆H₅—NO₂	5.7	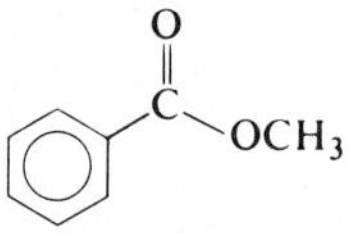 C₆H₅—C(=O)—OCH₃	−12
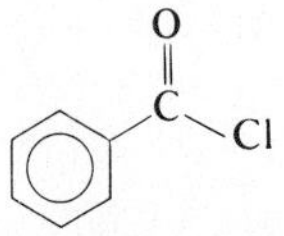 C₆H₅—C(=O)—Cl	−1	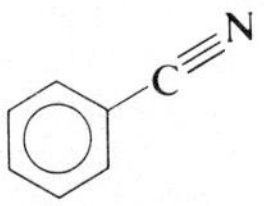 C₆H₅—C≡N	−13
CH₃CH₂CH₃	−190	CH₃—O—CH₃	−138
CH₃—S—CH₃	−83	CH₃—SO—CH₃	18
CH₃—SO₂—CH₃	109		

2.3 Melting Points

The very strong intermolecular hydrogen-bonding forces that lead to high boiling points also lead to high melting points. Comparison of melting points of acetamide, N-methylacetamide, and N,N-dimethylacetamide strikingly shows the importance of hydrogen bonding. In

$$CH_3\overset{\overset{O}{\|}}{C}-NH_2 \qquad CH_3-\overset{\overset{O}{\|}}{C}-NH-CH_3 \qquad CH_3-\overset{\overset{O}{\|}}{C}-N(CH_3)_2$$

mp 82°   mp 28°   mp −20°

general, functional groups that contain —OH and —NH bonds exhibit strong intermolecular hydrogen bonding. The melting points of the pairs of compounds listed in Table 2-6 illustrate the importance of hydrogen

Table 2-6

MELTING POINTS OF COMPOUNDS THAT CONTAIN ACIDIC HYDROGENS AND THEIR ALKYLATED ANALOGS

Compound	mp,°	Alkylated analog	mp,°
$(CH_3)_3COH$	26	$(CH_3)_3COCH_3$	−109
C_6H_5—COOH	122	C_6H_5—COOCH$_3$	−12
$(C_6H_5)_2NH$	53	$(C_6H_5)_2NCH_3$	−7
$(C_6H_5)_2C=N-OH$	144	$(C_6H_5)_2C=N-OCH_3$	102
CH_3CH_2SH	−112	$CH_3CH_2CH_2SCH_3$	−113

bonding with certain functional groups. Notice that the —SH group does not hydrogen bond strongly even though R—SH compounds are more acidic than ROH compounds. The boiling points of ethyl mercaptan (CH_3CH_2SH, bp 37°) and dimethyl sulfide (CH_3SCH_3, bp 37°) substantiate the lack of importance of hydrogen bonding with —SH groups.

Let us again consider the pair of *cis-trans* isomers of 1,2-dichloroethene. Which isomer should have the higher melting point? Even though the *cis* isomer has an overall dipole moment, the *trans* isomer does have two internal dipoles which happen to cancel. It is thus difficult to make a prediction based on dipole moments. Symmetry, however, clearly predicts that the *trans* isomer should have the higher melting point. Indeed, it is found that the isomer which we called "trans" based on its boiling point melts at −50°, whereas the other isomer melts at −81°. From the study of the melting points of many *cis-trans* pairs of isomers, it is concluded that in general, the *trans* isomer possesses the higher melting point.

It is hoped that by the study of the correlation of boiling and melting points with molecular structure, the student gains the feeling that molecules are real things and not just imagined, abstract concepts. The weight, size, shape, and flexibility of a molecule plus any intermolecular forces such as dipole-dipole attractions and hydrogen bonding determine the physical properties of the compound in a very reasonable fashion.

2.4 SOLUBILITY

The solubility of organic compounds in various solvents is another physical property of organic compounds that is highly dependent on the kind, number, and environment of functional groups that the compound possesses. For example, many organic compounds commonly found on a dinner table, such as grease from meat, salad oil, and butter, are not significantly soluble in water, yet others, such as vinegar (acetic acid), sugar, and alcohol (ethyl, that is), are very soluble in water. A study of the molecular structures of these substances indicates that the water insoluble materials are composed of molecules that have large hydrocarbon portions, whereas the water soluble materials are composed of molecules whose structures are dominated by polar functional groups.

When a substance, called a solute, is dissolved in a solvent, molecules of it go from the condensed phase (solid or liquid), where they are surrounded by solute molecules, into the solution phase, where they are surrounded by solvent molecules. In discussing solubility, then, two states are important: (1) the condensed solute state and (2) the solution state. If the molecules of a substance are more stable in the condensed solute phase than the solution phase, then the substance will not be very soluble. On the other hand, a substance that is very soluble will be composed of molecules that are more stable in the solution phase than in the solute phase. If the structure of a solute molecule is changed in such a fashion that its interactions with solute molecules become more favorable but its interactions with solvent molecules are unaffected, then the solubility of the solute will decrease. Of course, structural changes that lead to more favorable interactions of solute molecules with solvent molecules rather than solute molecules will increase the solubility of the solute. Molecular alterations that lead to unfavorable interactions of the solute molecules with solute or solvent molecules will have the opposite effects on solubility. In considering the effect of a change of molecular structure on solubility, then, one must consider the effect of the change on the stability of the molecule in both states.

The solubility of a substance, A, can be expressed by Eq. 2.1 where A_{solute} and A_{solution} represent the concentration of A in the respective

$$A_{\text{solute}} \xrightleftharpoons{K_{\text{sol}}} A_{\text{solution}} \tag{2.1}$$

states at equilibrium. These concentrations will be related, of course, by the equilibrium constant, K_{sol}. Since the concentrations of organic mole-

2.4 Solubility

cules in the condensed phase for all organic compounds is about the same, the change in K_{sol} will be approximately proportional to the change in $A_{solution}$ as given by Eq. 2.2, where C_1 is a constant. Consequently an in-

$$\Delta K_{sol} = \left| \Delta \left(\frac{A_{solution}}{A_{solute}} \right) \right| \cong (C_1) \Delta A_{solution} \tag{2.2}$$

crease in the solubility of a substance A means an increase in $A_{solution}$, which means that K_{sol} increases. For convenience, when we want to describe the solubility of a substance in quantitative terms, we will give its solubility in moles/liter of solvent.

Let us first consider the solubility of organic compounds in water. The two states of importance are the condensed organic phase, where solute molecules will be surrounded by organic (solute) molecules, and the solution state, where solute molecules will be surrounded by water molecules. In the condensed organic phase, various intermolecular forces such as dipole-dipole attractions and hydrogen bonding will exist. In water, hydrogen bonding is the dominant intermolecular force. Thus any acidic (proton donating) or basic (proton accepting) portions of a solute molecule will most likely participate in this hydrogen bonding scheme when it enters the solution phase. Obviously, a hydrocarbon molecule (e.g., hexane, cyclohexene, or benzene) will not be very stable in an aqueous solution, since the only way it can participate in the highly structured hydrogen-bonding scheme of water is by *breaking up* strong hydrogen bonds between water molecules. This highly unstable solution state results in low solubility for most hydrocarbons. On the other hand, organic molecules that are loaded with hydrogen-bonding functional groups, such as sugar molecules which have many hydroxy groups, are quite stable in the water medium, since they readily form strong and numerous hydrogen bonds with water molecules.

$$\begin{array}{c} CH{=}O \\ | \\ CHOH \\ | \\ CHOH \\ | \\ CHOH \\ | \\ CHOH \\ | \\ CH_2OH \end{array}$$

a typical sugar

The effect of the addition of a hydroxy group to a molecule on the solubility of the compound in water is more clearly seen by looking at the solubility of hexane, 1-hexanol, and 1,6-hexanediol. At room temperature only 0.0016 mole of hexane will dissolve in a liter of water, but 1-hexanol is 30 times more soluble than hexane and 1,6-hexanediol is very soluble in water.

Although hydroxy groups increase the solubility of organic compounds by both accepting and donating hydrogen bonds, functional groups that can only accept hydrogen bonds also increase the solubility of organic compounds. Of course, the more polar functional groups should be more stable in the water medium than the less polar groups. Thus, N,N-dimethylaniline, $C_6H_5N(CH_3)_2$ is not soluble in water at room temperature but nitrobenzene ($C_6H_5NO_2$) is slightly soluble (0.016 mole/liter of water). This must be a result of the greater polarity of the nitro group. However, aniline, $C_6H_5NH_2$, which is capable of donating hydrogen bonds, unlike its N,N-dimethyl derivative, is 23 times more soluble than nitrobenzene.

Another functional group that increases the water solubility of an organic compound is the carboxy group. For example, at room temperature, benzoic acid, C_6H_5COOH, is slightly soluble in water (0.022 mole/liter of water). The carbomethoxy group, $—COOCH_3$, being polar and an acceptor of hydrogen bonds, would be expected to increase the solubility of an organic molecule and, indeed, methyl benzoate, $C_6H_5COOCH_3$, is slightly soluble in water at 30° (0.00115 mole/liter of water). However, it is interesting to note that benzoic acid, which is a hydrogen donor, is much more soluble than its methyl ester.

In summary, then, the solubility of an organic compound in water will be increased by the addition of a polar functional group. The most effective water-solubilizing functional groups are those that are capable of donating hydrogen bonds, such as hydroxy and carboxy groups.

Let us now consider the solubility of organic compounds in organic solvents. The properties of organic solvents can vary tremendously. For example, methanol would be expected to be similar to water since it is polar and capable of hydrogen bonding. On the other extreme, an alkane such as hexane should be very different from water since it is not polar and it cannot participate in hydrogen bonding either as a donor or as an acceptor. Naturally, organic solvents that fall in between these two extremes, such as 1-hexanol, should and do have intermediate properties. With respect to these three solvents, an organic compound that possesses polar functional groups should be most stable in methanol, and least stable in hexane. Thus it is not surprising that the solubility of sugar is greatest in methanol, least in hexane, and intermediate in hexanol.

In nonpolar organic solvents, then, the solubility of an organic compound will be decreased by the addition of a polar functional group. Also, the solubility of an organic compound that possesses a polar functional group will increase if the polarity of the organic solvent is increased.

With respect to the intermediate cases, generalizations are very difficult to make, since the balance of forces is very delicate and many factors are involved.

In conclusion, the old rule "like dissolves like" is still one of the best for predicting the solubility of organic compounds. Solvents that possess functional groups identical or similar to those of the solutes usually dissolve large amounts of the solute.

2.5 GAS-LIQUID PARTITION CHROMATOGRAPHY

An extremely important recently discovered tool of the organic chemist is gas-liquid partition chromatography (glpc), which is often referred to as gas chromatography (gc), vapor phase chromatography (vpc), or gas-liquid chromatography (glc). Glpc is used to separate mixtures of organic compounds quantitatively and is based on many of the principles that we have considered with respect to boiling points and solubility. Many excellent instruments are available today, and all consist essentially of the parts shown in Fig. 2.3. The injector is a device that enables one to

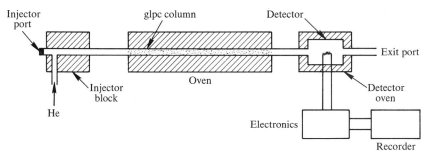

Fig. 2-3 Block diagram of a gas-liquid partition chromatography instrument.

introduce a mixture of compounds into a flow of helium which carries the mixture through a glpc column, a detector, and out. The glpc column is packed with some stationary solid phase (often firebrick particles) which is coated with some stationary liquid phase. A schematic diagram of a longitudinal section of a glpc column is shown in Fig. 2.4. The injector block, column, and detector are kept at some temperature by means of separate heaters and ovens. The injector port usually consists of a rubber stopper through which a syringe needle can be inserted in order to inject the sample into the flow of helium. If one injects a mixture of two organic compounds, say A and B, the heat of the injector block will vaporize these compounds and the flow of helium will carry them to the column. When the mixture arrives at the column, most of it will dissolve in the stationary liquid phase. However, both A and B will have a certain vapor pressure so that the gas (helium) in contact with the stationary liquid phase that

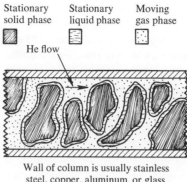

Fig. 2-4 Longitudinal section of a glpc column.

contains A and B will contain a certain amount of A and B. As the gas moves, some of the vapors of A and B that it contains will dissolve in the stationary liquid phase that contains no A and B; moreover, some of the A and B dissolved in the initial portion of the stationary liquid phase will vaporize into the fresh helium. Obviously, the speed with which A and B leave the stationary liquid phase depends on the vapor pressure of A and B. If A is more volatile than B, it will tend to leave the stationary liquid phase faster than B and will thus move through the column at a faster rate. Figure 2-5 illustrates typical positions of A and B as they move through the column.

When a compound leaves the glpc column, it rapidly passes through the detector which detects the presence of the compound. There are sev-

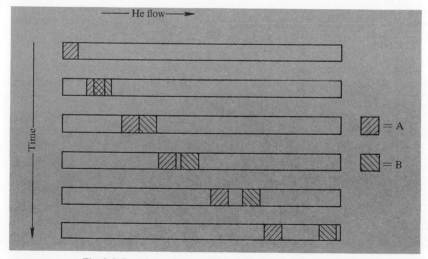

Fig. 2-5 Positions of two compounds, A and B, moving through a glpc column.

2.5 Gas-Liquid Partition Chromatography

eral different kinds of detectors. Among them are thermal conductivity detectors, which are widely used. Thermal conductivity detectors contain a device known as a thermistor, an electrical resistor whose resistance changes linearly with temperature. The thermistor is so placed in the detector that heat is transmitted from it to the walls of the detector through the flowing gas. The normal flow of helium will transfer a constant amount of heat from the thermistor and thus it will have a constant resistance. The thermistor is part of a Wheatstone bridge which is connected through an amplifier to a potentiometric strip-chart recorder. Initially, the thermistor's resistance remains constant and the Wheatstone bridge is balanced, which results in a straight base line by the recorder. When an organic compound flows through the detector, it conducts more heat from the thermistor. The resistance of the thermistor changes and this unbalances the bridge, which produces a voltage proportional to the imbalance. This voltage is amplified and fed to the recorder which produces a peak. Thus, one obtains a series of peaks as a function of time with each

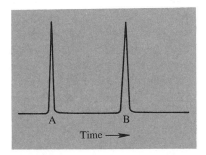

compound represented by a peak. The time it takes for a substance to go through the column is called its *retention time*. Of course, the retention time is a function of the particular column, the column's temperature, and the rate of helium flow.

Many different liquids have been used for the stationary phase and separations of amazingly similar compounds have been achieved. However, quite commonly a nonpolar high boiling liquid is used and, in these cases, the relative vapor pressures of the organic compounds correspond to the relative vapor pressures of the pure compounds. In other words, factors which affect the boiling point of a compound will affect its vapor pressure when it is dissolved in a nonpolar liquid in a similar fashion. Thus, the lower boiling compounds come through the column before the higher boiling compounds. Compounds that boil even a few degrees apart can be cleanly separated. If the glpc column contains a polar liquid phase, say a polyester which is a long chain compound that contains an ester functional group at regular intervals, it will interact with polar compounds to a greater extent and thus lower their vapor pressure. In other words, the more polar compound is more soluble in the polar liquid phase. This then causes the retention time of a polar compound such as a ketone to be much higher than that of a nonpolar compound, even though the nonpolar compound might have the higher boiling point. Thus a carbowax 20M column, a poly(ethylene glycol), will permit decalin (bp 186°) to pass through much faster than acetone (bp 56°).

Glpc is used to analyze mixtures of compounds both qualitatively and quantitatively. If an unknown compound has the same retention on a given column under a given set of conditions as a known compound, K, it is possible that the unknown compound is K. However, identical retention times are not rigorous proof of identity, since many compounds, especially isomers, may have the same retention times. If the unknown compound has the same retention time as K on several very different columns, then it is much more likely that the unknown is K. However, comparison of glpc retention times can never unequivocally prove the structure of a compound. A more nearly rigorous proof of structure is obtained by collecting the compound from the effluent helium in a small condenser and taking infrared, nuclear magnetic resonance, and mass spectra of it.

Since the area of a peak is proportional to the amount of compound that passes through the detector, glpc can easily be used for quantitative analysis. For example, if a mixture contains two compounds and the glpc peak of one of the compounds is twice that of the other, then it is *approximately* a 2:1 mixture. The mixture can only be analyzed approximately unless one knows the relative thermal conductivities of each compound, since each compound conducts heat at a different rate. Often the relative thermal conductivity of two compounds is close to 1.0 but values of 2.0 are not uncommon.

2.6 ULTRAVIOLET-VISIBLE (ELECTRONIC), INFRARED, AND NUCLEAR MAGNETIC RESONANCE SPECTROSCOPY

When any kind of electromagnetic radiation is absorbed by a molecule, the molecule enters a higher energy state. Usually the molecule dissipates this energy rapidly by collisons with other molecules so that one molecule does not remain in an excited state for any length of time. The distributed energy shows up as translational energy which, of course, causes the material to become warm.

Absorption of electromagnetic radiation is quantized, which means that a given quantity of energy is transferred to the molecule essentially instantaneously. The particular energy transition, i.e., the difference in energy between the ground state and the excited state, will depend on the amount of energy absorbed. The energy absorbed is related to the frequency of the absorbed radiation by

$$E = h\nu \qquad (2.3)$$

where E is the energy, h is Planck's constant, and ν is the frequency of radiation. Since ν is proportional to $1/\lambda$ where λ is the wavelength of the radiation, Eq. (2.3) shows that the energy of electromagnetic radiation is inversely proportional to its wavelength.

Three important types of transitions take place: (1) electronic, (2) vibrational, and (3) rotational. It is found that electronic transitions require the most energy, vibrational transitions less energy, and rotational transitions even less energy. Thus, as shown in Fig. 2-6, as the frequency of radiation decreases, one goes from electronic to vibrational to rotational transitions.

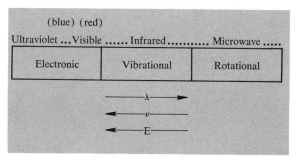

Fig. 2-6 Main types of molecular excitations at various wavelengths of electromagnetic radiation.

A. Ultraviolet-Visible Spectroscopy: The absorption of visible electromagnetic radiation leads, of course, to color. The color of a substance is a result of the light shining on it minus the color of light it absorbs. Thus, in white light a white substance absorbs no light, a black substance absorbs all colors, a yellow substance absorbs blue, a green substance absorbs orange, a purple substance absorbs green-yellow, etc. The electronic (ultraviolet-visible) spectrum of a substance is measured by passing light of various wavelengths through a sample of the substance and measuring the amount of light absorbed. The relationship $\log(I_0/I)$ where I_0 is the intensity of the incident light and I is the intensity of the transmitted light gives the absorbance A or optical density O.D. A typical spectrum is a plot of absorbance A against wavelength λ, in millimicrons (mμ). An example of such a plot is shown in Fig. 2-7. There are two important features of these spectra: the positions and heights of the maxima. The position of a maximum is referred to as a "λ_{max}." A spectrum will have as many λ_{max}'s as peaks. The height of the peak is given by the molar extinction coefficient, ϵ, which is related to the absorbance by Eq. (2.4), where c is the molar concentration of the substance and l is the path length

$$\epsilon = \frac{A}{cl} \qquad (2.4)$$

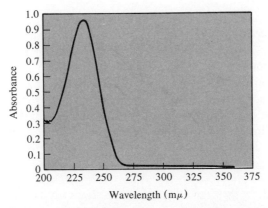

Fig. 2-7 Ultraviolet absorption spectrum of a typical α,β-unsaturated ketone, 3,6,6-trimethylcyclohex-2-enone in ethanol, concentration = 10^{-4} M. (From A. Ault, *Problems in Organic Structure Determination*, McGraw-Hill, 1967, p. 104.)

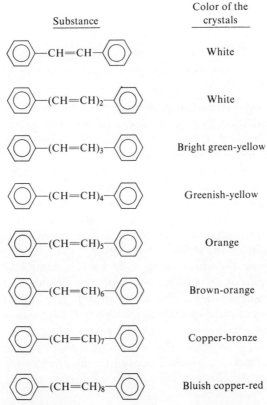

Fig. 2-8 Color of the crystals of conjugated α,ω-diphenylpolyenes.

2.6 Ultraviolet-Visible, Infrared, and Nuclear Magnetic Resonance Spectroscopy

of the solution in centimeters. Thus, it is customary to report the λ_{max} and ϵ for each major peak of the electronic spectrum of a substance. As seen in Fig. 2-6, the shorter wavelength, higher energy visible radiation is blue, and the longer wavelength, lower energy visible radiation is red. Thus, an electronic excitation that requires much energy absorbs blue light, whereas an electronic excitation that requires less energy absorbs red light. Higher energy transitions, which are actually very common, absorb ultraviolet radiation, which is below the visible region. Study of the color of the crystals of the compounds given in Fig. 2-8 suggests that there must be some correlation between the number of conjugated double bonds and the energy necessary to bring about an electronic excitation. In order for us to understand the effect of structure on the absorption of visible light, we must consider the electronic structure of molecules in more detail.

During an electronic transition, the electrons are excited from the ground state to some higher energy electronic state. It is convenient to think in terms of molecular orbitals when talking about molecular electronic excitations. Just as atomic orbitals describe the position of electrons around one nucleus, molecular orbitals describe the position of electrons around two or more nuclei as shown in Fig. 2-9.

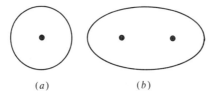

Fig. 2-9 (a) An atomic orbital, (b) a molecular orbital.

The hydrogen atom can be thought of as a stationary nucleus with an electron revolving around it. Since the electron is moving, it possesses kinetic energy and since there is an electrostatic interaction between the two particles, it possesses potential energy. *Wave mechanics* treats the electron as a standing wave and a *wave equation* can be written which describes this situation. This wave equation is a differential equation and acceptable roots to it are called *wave functions* or *eigenfunctions* and are often represented by φ's or ψ's. A wave function is called an *orbital* and is a function of x, y, and z and has a finite value throughout all space. How-

$$\psi = f(x, y, z)$$

ever, its value is very small at a distance of more than a few Ångstrom units from the nucleus. The value of the square of the wave function at any point is proportional to the probability of finding an electron at that point. Probability maps of finding an electron around a nucleus give rise to the pictures of orbitals.

A particular energy is associated with each wave function. For the hydrogen atom, the lowest energy orbital is called the 1s orbital, the next higher energy orbital is the 2s, the next higher energy orbital is the 2p, etc.

Polyelectronic atoms cannot be easily treated exactly because one electron will influence the others and thus all particles must be considered. Molecules are even more difficult to treat exactly since several nuclei must be considered. For polyelectronic atoms, one obtains a set of orbitals (we will not worry at this time how this set is obtained) and feeds in the proper number of electrons starting with the lowest energy orbitals according to Pauli's principle and Hund's rules. This is called the *aufbau principle* and leads to the ground state electronic configuration. An excited state configuration will have one or sometimes two electrons in a higher orbital than they occupied in the ground state. For example, the ground state and an excited state of sodium are shown in Fig. 2-10. The amount of energy needed to reach this excited state is approximately equal to the difference in energy levels (ΔE in Fig. 2-10) between the two orbitals. This energy

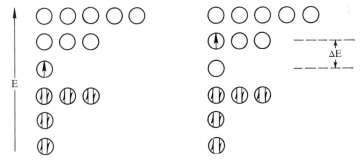

Fig. 2-10 Ground and excited state configurations for a typical atom, sodium.

difference is only approximately equal to the difference in energy between the ground and excited state, since the orbitals in the excited state change energy slightly due to the removal of the electron from the lower orbital. Of course, there are many excited states since in theory any electron can be moved to any available orbital. Thus the electronic spectrum of an atom consists of several lines, each corresponding to a discrete energy transition from one state to another. The absorption spectrum of sodium is shown in Fig. 2-11.

The electronic spectra of molecules can be considered in exactly the same fashion as those of polyelectronic atoms except that one starts with a set of molecular orbitals instead of atomic orbitals. An approximation used to generate a set of molecular orbitals is to take a linear combination of atomic orbitals (LCAO). Consideration of the π-bond of ethylene

2.6 Ultraviolet-Visible, Infrared, and Nuclear Magnetic Resonance Spectroscopy

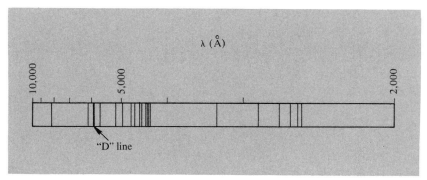

Fig. 2-11 Ultraviolet-visible absorption spectrum of sodium.

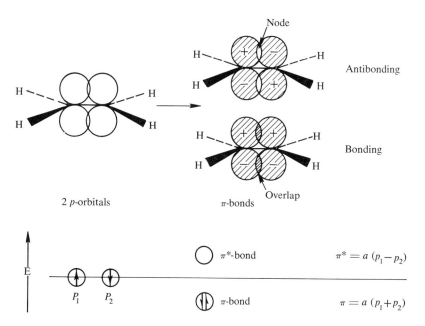

illustrates this method. The π-bond of ethylene is a molecular orbital since it covers more than one nucleus. A way of generating a wave function that describes this π-bond is to take a linear combination of the two p-orbitals on the carbon atoms. In other words, the π-bond can be thought of as being made up of $\frac{1}{2}$ of each p-orbital. The other $\frac{1}{2}$ of each p-orbital makes up another molecular orbital called the π^*, or antibonding, orbital (the star is often used to represent antibonding orbitals). Thus, mathematical combination of the two p-orbitals gives rise to the two molecular orbitals π and π^*. These can be represented by Eqs. (2.5) and (2.6), where

$$\pi^* = a(p_1 - p_2) \qquad (2.5)$$

$$\pi = a(p_1 + p_2) \tag{2.6}$$

where p_1 and p_2 are wave functions for the two atomic p-orbitals and a is a constant which adjusts the scale of the MO's. In the bonding orbital π the atomic orbitals have the same sign and consequently there is constructive interference where they overlap. On the other hand, in the antibonding orbital π^* the atomic orbitals have opposite signs and therefore destructive interference, which results in a node, occurs where they overlap. This constructive and destructive interference is represented respectively by heavily shaded and unshaded areas in the above diagram.

In a fashion similar to the generation of π- and π^*-bonds, σ- and σ^*-bonds can be generated. In Fig. 2-12 are shown the electronic configurations of the ground state and an excited state for benzene. Just as with polyelectronic atoms, the energy needed to reach this excited state is approximately given by ΔE.

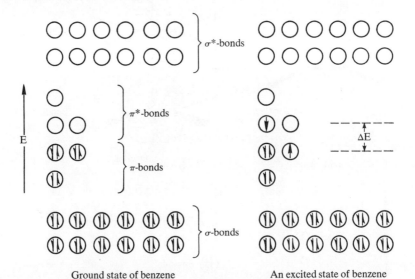

Fig. 2-12 Electronic configurations of ground and excited states of a typical molecule, benzene.

The absorption spectrum of benzene (Fig. 2-13), like that of other molecules, is not a series of sharp lines but is composed of rather broad bands. This is a result of the existence of vibrational and rotational energy levels between the electronic transitions† (Fig. 2-14), i.e., the exact energy difference between the ground and excited states will depend upon what vibrational and rotational states the molecule is in both in the ground and excited states.

† N. L. Allinger and J. Allinger, *Structure of Organic Molecules*, in this series, p. 58, Fig. 4-6.

2.6 Ultraviolet-Visible, Infrared, and Nuclear Magnetic Resonance Spectroscopy

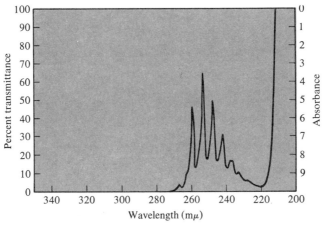

Fig. 2-13 Ultraviolet absorption spectrum of benzene in 95% ethanol, concentration = 2.92×10^{-3} M. (From D. J. Pasto and C. R. Johnson, *Organic Structure Determination*, Prentice-Hall, 1969, p. 106.)

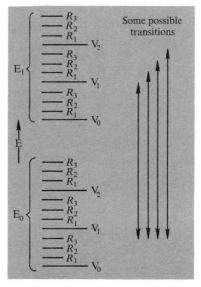

Fig. 2-14 Electronic, vibrational, and rotational energy levels of a typical molecule.

In general, if a molecule contains filled π-orbitals, these will be the highest energy filled orbitals. Moreover, the lowest unoccupied orbitals will be π^*-orbitals. In other words, the grouping of the σ- and π-orbitals for benzene, shown in Fig. 2-12, is the situation that is found for most molecules that contain π-orbitals. This means that the lowest energy electronic transitions will usually involve π- and π^*-bonds and consideration of these orbitals alone helps one understand much about electronic excitation. Excitations of this type are referred to as $\pi \to \pi^*$ transitions.

The Hückel molecular orbital method is an approximate method used to calculate π molecular orbitals and the energies of these MO's from a linear combination of atomic p-orbitals. These MO's will extend over the whole π-system. Thus four adjacent p-orbitals give rise to MO's over four-atoms, etc. This method and more sophisticated molecular orbital methods lead to the conclusion that the lowest energy MO's have the least number of nodes. In Fig. 2-15, the MO's for ethene, 1,3-butadiene, and

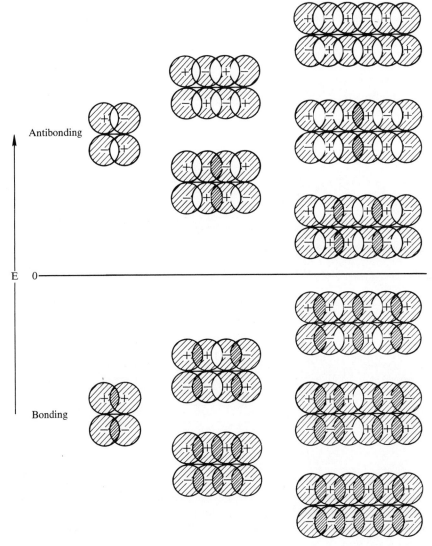

Fig. 2-15 Molecular orbitals for ethene, 1,3-butadiene, and 1,3,5-hexatriene.

1,3,5-hexatriene are depicted. Notice that the wave function ψ changes sign whenever a node is crossed. It is not surprising that the lowest energy orbitals have the least number of nodes since with all waves the longest ones possess the least energy. The lowest energy state of a vibrating string is its fundamental frequency, which has no nodes.

Another reasonable characteristic of the sets of MO's shown in Fig. 2-15 is that the orbitals get closer together in energy as the number of atoms increases. This is also not surprising since the addition of an extra node to a longer wave always requires less energy than the addition of an extra node to a shorter wave.

A further discussion of molecular orbitals is to be found in L. M. Stock's book in this series, *Aromatic Substitution Reactions*.

Since the main electronic transition for a π-system is from the highest occupied MO to the lowest unoccupied MO, it is easy to see that as a π-system gets bigger, the transitions require less energy. The λ_{max} for the systems shown in Fig. 2-16 clearly illustrate this point.

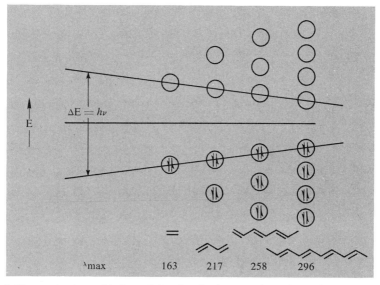

Fig. 2-16 π-molecular orbitals and λ_{max}'s of ethene and conjugated polyenes.

We have considered the position of the λ_{max}, but have said nothing about ϵ, the magnitude of the absorption. The value of ϵ is governed by a set of rules called *selection rules*. The probability that a certain transition can occur is given by these selection rules. The higher the probability, the greater ϵ will be. A *forbidden transition* is one that has a low probability of occurring. These selection rules are complicated and will not be discussed here. However, the student should remember that similar transitions usually have similar ϵ's. Thus, the longest wavelength transitions given in Fig. 2-16 for the series of polyolefins are all the same type of transition and thus all have ϵ's of the same order of magnitude.

In addition to $\pi \to \pi^*$ transitions, $\sigma \to \sigma^*$ transitions and $n \to \pi^*$ transition are known. In a $\sigma \to \sigma^*$ transition, an electron in a bonding σ-orbital goes to an antibonding σ-orbital, and in an $n \to \pi^*$ transition, an electron in a nonbonding orbital goes to an antibonding π-orbital. The $n \to \pi^*$ transition is common with ketones which have two nonbonding pairs of electrons on the oxygen atom. These three types of transitions are illustrated in Fig. 2-17.

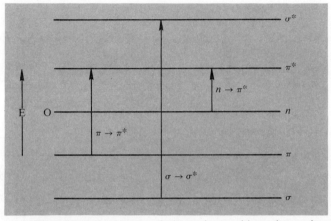

Fig. 2-17 Various types of electronic transitions that carbonyl groups undergo.

Any functional group that absorbs ultraviolet-visible radiation is called a *chromophore*. An olefin is certainly one of the most important chromophores, but other functional groups, especially those that contain π-systems, are also very important. Some important chromophores are

$$\begin{array}{c}\diagdown\\ \diagup\end{array}\!\!C\!=\!O, -C\!\equiv\!C\!-, -COOH, -C\!\equiv\!N, -NO_2, -N\!=\!N\!-, \text{ and } =\!N_2.$$

These chromophores have characteristic absorptions just as the olefins do and behave similarly inasmuch as longer λ_{max}'s are observed when they are attached to longer π-systems. The series of aldehydes shown in Fig. 2-18 illustrates this point.

Aldehyde	λ_{max}	ϵ
$CH_3-CH=O$	180	10,000
$CH_3-CH=CH-CH=O$	217	15,650
$CH_3-CH=CH-CH=CH-CH=O$	270	27,000
$CH_3-CH=CH-CH=CH-CH=CH-CH=O$	312	40,000

Fig. 2-18 λ_{max}'s and ϵ's of acetaldehyde and unsaturated aldehydes.

2.6 Ultraviolet-Visible, Infrared, and Nuclear Magnetic Resonance Spectroscopy

Auxochromes are groups that do not have any significant absorption above 200 mμ by themselves, but change the λ_{max} of a chromophore significantly when attached to it. These groups usually have nonbonding electrons that can interact with the π-system of the chromophore. Some common auxochromes are —OH, —Cl, —Br, and —NH$_2$. The longest wavelengths λ_{max} for the following aromatic systems show the effect of auxochromes: C_6H_6, 254; C_6H_5OH, 270; C_6H_5Cl, 264; C_6H_5Br, 261; and $C_6H_5NH_2$, 280.

There are several ways that one can use the electronic spectrum of a substance for structural information. First, in a series of compounds, the compound with the longest π-system would have the highest λ_{max}. Thus the two tetraenes **1** and **2** can be readily distinguished. Second, all com-

1   **2**

pounds with a similar π-system will have similar spectra. Thus azulenes and naphthalenes can be distinguished easily. A more subtle use of electronic spectra for structure determination is to distinguish between *cis-*

$C_{10}H_8$

$_{max}(\epsilon)$: 221(110,000), 276(5,750), 286(3,900), 311(240)

$C_{10}H_8$

$\lambda_{max}(\epsilon)$: 274(61,500), 340(4,700), 579(324), 631(324), 659(150), 665(140)

and *trans*-stilbene. The two phenyl groups in the *cis* isomer would be

λ_{max}: 294
ϵ: 24,000
mp: 124°

λ_{max}: 278
ϵ: 9,350
mp: −5°

expected to interact and twist each other out of the plane of the olefin π-system. Thus, the longer π-system of the *trans* isomer should and does give rise to the higher and more intense λ_{max}. Notice that melting points confirm this assignment.

The presence of certain functional groups can be determined by electronic spectra. For example, from the electronic spectrum of a compound, one can tell if the compound contains a carbonyl group and whether or not it is conjugated. However, this type of information is often much easier to obtain from the infrared spectrum of the compound, which is usually more conclusive anyway. Nevertheless, the electronic spectrum is used as confirmatory information and often gives information about the major structure of the π-system which would be difficult to obtain otherwise.

B. Infrared Spectroscopy: The absorption of infrared radiation leads to vibrational excitations. Most molecules at room temperature are in their lowest vibrational state. If we consider a diatomic molecule, A—B, this means that A is oscillating with respect to B at some natural frequency, just as two balls attached by a spring might oscillate if the spring is stretched and then released. If the A—B bond has a dipole that changes as the bond vibrates and the frequency of electromagnetic radiation is equal to the frequency of the vibration of the bond, the radiation is absorbed and the molecule enters the higher vibrational state. An understanding of this absorption process is obtained if one considers the electrical portion of the electromagnetic radiation. In the presence of electromagnetic radiation, the molecule A—B may be considered as being in an oscillating electrical field. Since the bond has a dipole, the oscillating electric field will alternately attract and repel the atoms. If both the field and the atoms oscillate at the same frequency, absorption of energy takes place and the amplitude of the oscillation of the atoms increases. The molecule soon loses energy through collisions and drops to the lowest vibrational state. Of course, this dissipated energy shows up as heat (translational energy) and this process is the reason infrared lamps are used to cook food and keep it warm.

$$
\begin{array}{ccc}
| & \delta^- \leftrightarrow \delta^+ & | \\
|\,+ & \text{A—B} & -\,| \\
& \leftarrow \;\;\; \rightarrow & \\
\\
|\,- & \text{A—B} & +\,| \\
& \rightarrow \;\;\; \leftarrow & \\
\\
|\,+ & \text{A—B} & -\,| \\
& \leftarrow \;\;\; \rightarrow & \\
\end{array}
$$

etc.

Usually the region from 2.5–15 μ is used for organic compounds since this region contains most of the frequencies of interest. It is customary to plot % transmittance T against wavelength in microns (see Fig. 2-19). The % transmittance T equals I/I_o where I_o is the intensity of the incident light and I is the intensity of the transmitted light. Since $E = h\nu$, it is more convenient to think of absorptions in terms of frequency rather than wavelength, since the former is directly proportional to the energy. The frequency unit customarily used with respect to infrared absorption is reciprocal centimeters, cm^{-1}, which is simply the reciprocal of the wavelength in centimeters. Although organic chemists have long used the wavelength unit, microns, the trend is now toward using reciprocal centimeters.

There are three important aspects of infrared spectroscopy for the organic chemist. First, and by far the most important aspect, is that *many functional groups have characteristic infrared absorption frequencies.* Second, the exact position of this frequency can give information about the structural environment of the functional group. Third, the infrared spectrum can serve as a *fingerprint* to identify the compound since infrared spectra are complicated and contain many bands with some of them being unusually shaped.

The carbonyl group

$$\diagdown C = O$$

serves to illustrate the principle of characteristic absorptions. Almost any compound that contains a carbonyl group will show a strong infrared absorption around 1700 cm^{-1}. As seen in Fig. 2-19, cyclohexanone ab-

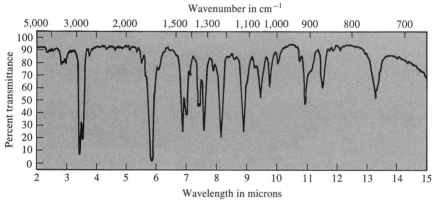

Fig. 2-19 Infrared spectrum of cyclohexanone. (From *An Introduction to Modern Experimental Chemistry* by R. M. Roberts, J. C. Gilbert, L. B. Rodewald, and A. S. Wingrove. Copyright © 1969 by Holt, Rinehart and Winston, Inc. Reprinted by permission of Holt, Rinehart and Winston, Inc.)

sorbs strongly at 1710 cm^{-1}. The presence or absence of a carbonyl group is almost definitely established by the presence or absence of this infrared band. But why should this principle hold? The stretching frequency of vibration of two atoms, A and B, attached by a covalent bond is given by Eq. (2.7), where ν is the frequency in cm^{-1}, c is the velocity of light, k is

$$\nu = \frac{1}{2\pi c} \sqrt{\frac{k}{M_A M_B / (M_A + M_B)}} \qquad (2.7)$$

the force constant of the bond in dynes/cm, and M_i is the atomic weight of atom i in grams. Now, when one of the atoms, say A, is attached to another atom, it turns out that the mass that is important is the mass of A, not the mass of A plus the rest of the molecule to which it is attached. Because of this, most bonds that have a dipole moment absorb infrared radiation at a characteristic frequency. The most characteristic frequencies of common bonds encountered by an organic chemist are given in Table 2-7. Notice that these are frequencies for bonds, not functional groups. Of course, functional groups are composed of one or more of these bonds, but the exact position of a bond's absorption might vary from one functional group to another. For example, a normal ketone absorbs at 1710 cm^{-1} but a normal ester absorbs at 1735 cm^{-1}. Eventually, the student should remember some of the frequencies for the more common functional groups, but at the onset, the student is advised to memorize the broad frequencies for common *bonds* given in Table 2-7.

Table 2-7

CHARACTERISTIC STRETCHING FREQUENCIES OF COMMON BONDS

Bond	Infrared band (cm^{-1})
C—H (aliphatic)	2850–2960
C—H (aromatic or olefinic)	3000–3100
O—H (not hydrogen bonded)	*ca.* 3600
O—H (hydrogen bonded)	3200–3600 (broad)
N—H	*ca.* 3500
C=O	1690–1760
C—O	1080–1300
C≡N	*ca.* 2200

The exact position of the frequencies can be very informative at times. For example, a normal ketone absorbs at 1710 cm^{-1}; however, an α,β-unsaturated ketone absorbs at 1690 cm^{-1}. The reason an α,β-unsaturated ketone absorbs at a lower frequency is that the C=O is weakened due to the following type of resonance which gives the C=O some single bond character. A weaker bond should vibrate at a lower frequency and

$$\left[\begin{array}{c} \diagdown \\ \diagup \end{array} C=C-C=O \longleftrightarrow \begin{array}{c} \diagdown \\ \diagup \end{array} \overset{+}{C}-C=C-O^- \right]$$

thus absorb lower frequency, lower energy radiation. Thus the structures of **3** and **4** are easily differentiated since **3** absorbs at 1690 cm^{-1} and **4** absorbs at 1710 cm^{-1}.

3   4

The infrared spectrum of a compound is also used as its "fingerprint" since a large number of irregularly shaped peaks makes most infrared spectra unique. Thus, if two samples have identical infrared spectra, it is quite likely that they are identical. However, one must be careful that they are not isomers with only subtle structural changes, since these often have nearly identical infrared spectra.

The student should work the problems at the end of this chapter in order to gain an appreciation for the use of infrared spectroscopy in organic chemistry.

A more extensive discussion of infrared spectroscopy can be found in J. R. Dyer's book in this series, *Applications of Absorption Spectroscopy of Organic Compounds*.

C. Nuclear Magnetic Resonance: The nuclei of many elements have a characteristic called *spin* associated with them. One might, for simplicity, think of the nucleus as being a hard sphere spinning on its axis. Some nuclei, like the hydrogen nucleus, can spin in two different directions. Normally, it doesn't make any difference whether the nucleus is spinning from east to west or from west to east. However, when the nucleus is placed in a strong magnetic field, it does matter in which direction the nucleus is spinning. Thus, it is easier for the nucleus to spin in one direction than the other and therefore one spin state is of lower energy than the other.

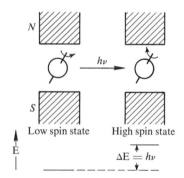

When protons in a magnetic field are exposed to electromagnetic radiation of a certain frequency, the phenomenon of resonance occurs, during which the radiation can be absorbed by the nucleus, causing it to go from the low spin state to the high spin state. The energy of the radiation that must be used for resonance to occur is directly proportional to the magnetic field in which the nucleus is. For example, if the strength of the field is doubled, the energy needed to "flip" the nucleus from one spin state to the other must be twice as great. For obvious reasons, this phenomenon is called *nuclear magnetic resonance* and is abbreviated NMR.

Nuclear magnetic resonance is of utmost importance to the organic chemist because the magnetic field that is important is not the applied magnetic field but the field the nucleus "feels." These are almost never the same because the electrons around the nucleus "shield" it in part from the applied field. Thus, the magnetic field at the nucleus is almost always less than the applied field.* One might compare the applied magnetic field to the sun and the electron cloud to real clouds. Of course, as clouds become thicker, less sun gets through. In a very similar fashion, as the electron density around a nucleus increases, less of the applied magnetic field gets to the nucleus. If the frequency of the electromagnetic radiation is kept fixed, then as the electron density around a nucleus increases, a greater applied magnetic field is needed for resonance (absorption of the radiation) to occur. Measurement of the strength of the applied field that is needed for absorption to occur tells one about the electron density around the nucleus which absorbs the radiation. Thus in an organic molecule that contains a respectable number of hydrogen atoms, the protons are little probes giving information about the density and, therefore, the shape of the electron cloud that envelopes the nuclear framework.

The inductive effect: Now let us look at what information is obtained from these probing nuclei. The information from NMR is given in the form of a spectrum shown in Fig. 2-20. Each peak corresponds to an absorption of radiation of a fixed frequency at a different applied magnetic field strength. In order to discuss the peak position quantitatively, it has been customary to talk about the shift downfield from a

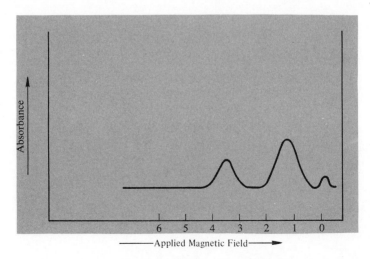

Fig. 2-20 Poorly resolved NMR spectrum of ethyl chloride.

---

* The electrons shield the nucleus since they tend to circulate in a plane perpendicular to the magnetic field in which they are placed. The circulation of these electrons, like any charges, creates a magnetic field in the opposite direction. Thus the induced field is subtracted from the applied field to give the resultant field to which the nucleus is exposed.

2.6 Ultraviolet-Visible, Infrared, and Nuclear Magnetic Resonance Spectroscopy

standard peak. The standard peak is the signal of the protons of tetramethylsilane, TMS. Since the four methyl groups around the silicon atom

$$\begin{array}{c} CH_3 \diagdown \quad \diagup CH_3 \\ CH_3 \blacktriangleright Si \\ | \\ CH_3 \end{array}$$

are arranged tetrahedrally, all the methyl groups are equivalent. Moreover, since all the methyl groups are rotating rapidly, all the hydrogen atoms are equivalent. Thus all twelve hydrogen nuclei have identical electron clouds around them and only one NMR absorption occurs for TMS. Now the NMR signal for ethane is also one peak, since all six hydrogen atoms are equivalent. But the peak is 1.1 parts per million (p.p.m.)* downfield from the TMS peak, which is symbolized as 1.1δ. Since this means that resonance occurs at lower field, the cloud of electrons around the hydrogen nuclei of ethane must be less dense than the cloud of electrons around those of TMS. In fact, TMS was chosen as a standard since relatively few substances have protons with more electron density around them than those of TMS.

Now let us look at the NMR spectrum of methyl chloride. Again the hydrogen atoms are all equivalent since they are symmetrically placed, but now the signal occurs at 3.4δ. The chlorine has decreased the electron density around the hydrogen atoms. This is just what is expected based on

$$CH_3 - Cl$$
$$\leftrightarrow$$

dipoles, since we know that methyl chloride has a dipole and that the negative end of the molecule is the chlorine side. The hydrogens then must be more electron poor than those of ethane and so give rise to an NMR signal at lower field. This polarization of electrons is called the *inductive effect*. In other words, chlorine, being more electronegative than a methyl group, withdraws electrons from a methyl group to a greater extent than another methyl group. The electron cloud so adjusts itself around the nuclei of methyl chloride to be lopsided in favor of the chlorine. Now what happens with ethyl chloride? The electronegative chlorine can still remove electronic charge from the adjacent hydrogen and then give rise to an NMR signal at about the same place as that of methyl chloride. However, since the chlorine is further removed from the methyl group of ethyl chloride, the methyl NMR peak is upfield almost to the same place as that for ethane (see Fig. 2-20). Thus, the inductive effect drops off fairly quickly with distance. Note that the inductive effect can be looked upon as a purely electrostatic phenomenon. This means that the electron cloud

---

* For this discussion, it is not necessary for the reader to know what these units of displacement of peak mean. It is only necessary for him to realize that as this number increases, the peak shifts further downfield from the TMS peak.

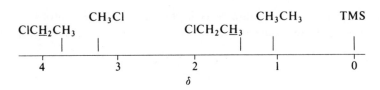

is being deformed to lower the electrostatic potential energy. In cases where there is a full positive charge, it is easy to see why the electrons are attracted to it. Thus the methyl groups of tetramethylammonium ion are

$$CH_3-N\begin{matrix}CH_3\\ \\CH_3\end{matrix} \qquad CH_3-\overset{CH_3}{\underset{CH_3}{\overset{|}{N}}}\!\!\overset{+}{-}CH_3$$

shifted downfield 1.1 p.p.m. from those of trimethylamine. In cases where there are no full charge separations, such as the alkyl halides, one can look at the electronegative atom as inducing a dipole along the bond which attaches it to the carbon atom. For ethyl chloride, the dipole along the carbon-chlorine bond effectively makes the α-carbon atom a little more positive and this, in turn, induces dipoles along the bonds of the atoms joined to it. This effect would naturally fall off with distance.

$$H\!\rightarrowtail\!\overset{\overset{H}{\downarrow}}{\underset{\underset{H}{\uparrow}}{C}}\!\rightarrowtail\!\overset{\overset{H}{\downarrow}}{\underset{\underset{H}{\uparrow}}{C}}\!\rightarrowtail\!Cl$$

In addition to the redistribution of charge along bonds, an electrostatic effect can be exerted directly through space if the molecule is not linear. This is called the *field effect* and can be neglected in this discussion since it is difficult to separate from the usual inductive effect.

The resonance effect: The inductive effect accounts very well for the electron density around hydrogens in saturated systems. The NMR spectrum of cyclohexanone shows that the electron density on the hydrogen atom decreases as the carbonyl group is approached. However in 2-cyclohexenone the electron density is greater on the α-hydrogen atom than on the β-hydrogen atom.* Moreover, in pyridine, the γ-hydrogen atom is more deshielded than the β-hydrogen atoms, even though the β-hydrogen atoms are closer to the electronegative nitrogen. Certainly in *p*-methoxytoluene the most electronegative atom is oxygen, yet the hydrogens *meta* to the methoxy group are more deshielded than those of the *ortho* hydro-

---

* Both the α- and β-hydrogens of 2-cyclohexenone are shifted much further downfield than those of cyclohexanone since they are olefinic.

2.6 Ultraviolet-Visible, Infrared, and Nuclear Magnetic Resonance Spectroscopy

$$\begin{array}{c} CH_2-CH_2 \\ / \quad \quad \backslash \\ CH_2 \quad \quad C=O \\ \backslash \quad \quad /\leftrightarrow \\ CH_2-CH_2 \end{array}$$

NMR signals ⟶   1.7   2.3   (p.p.m. from TMS)

$$\begin{array}{c} CH_2-CH_2 \\ / \quad \quad \backslash \\ CH_2 \quad \quad C=O \\ \backslash \quad \quad /\leftrightarrow \\ C=C \\ / \quad \backslash \\ H \quad \quad H \end{array}$$

NMR signals ⟶   7.00   5.85   (p.p.m. from TMS)

gen atoms. Some effect other than the inductive effect must be operating. Indeed, all of these seeming anomalies are easily accounted for by the

[pyridine structure with 7.60 δ, 7.00 δ, 8.60 δ]   [p-methylanisole structure with OCH₃, 6.80 δ, 7.05 δ, CH₃]

resonance* effect, which takes into account the wave nature of electrons. Wave mechanical calculations indicate that in a distorted π-system such as an α, β-unsaturated ketone, the effect of the distortion is felt in an

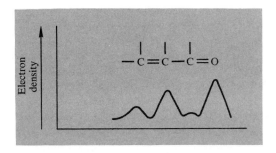

alternating fashion. The organic chemist is fortunate in having a very simple means of dealing with this alternating effect. The theory of resonance beautifully accounts for this uneven charge distribution in terms that are so simple that they seem almost artificial.

The theory of resonance tells us that the real wave function of a molecule is best represented by the linear combination of wave functions for resonance structures for that molecule. Good resonance structures are

---

* There is no relationship between the use of the word "resonance" here and in "nuclear magnetic resonance."

those that have few formal charges and the formal charges that do exist should be placed on atoms that can bear them well.² The lowest energy resonance structures will be the main contributors to the real structure of the molecule. Structures **A**, **B**, and **C** are the three best resonance

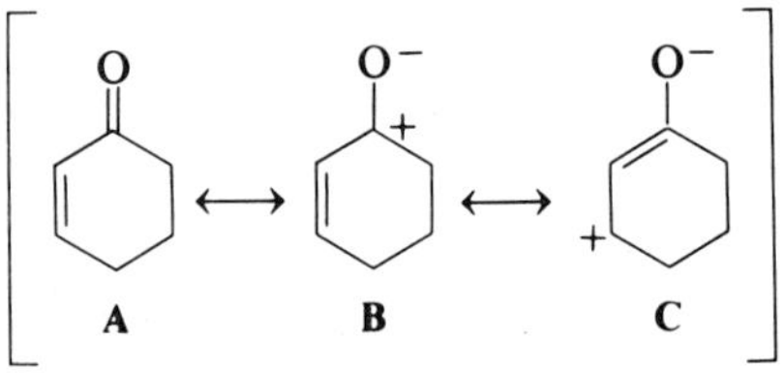

structures for 2-cyclohexenone. It is seen that one good resonance structure has a positive charge on the β-carbon atom but none has a positive charge on the α-carbon atom. Indeed, one cannot draw a good resonance structure with positive charge on the α-carbon atom. Thus, the resonance picture shows that the electron density of the π-system is less at the β-carbon atom than at the α-carbon atom. The NMR spectrum indicates that this greater deficiency of π-electrons at the β-carbon atom is enough to deshield the β-hydrogen atom more than the α-hydrogen atom even though the inductive effect deshields the α-hydrogen atom more. In a similar fashion, resonance accounts for the unexpected electron distribution in pyridine and *p*-methoxytoluene.

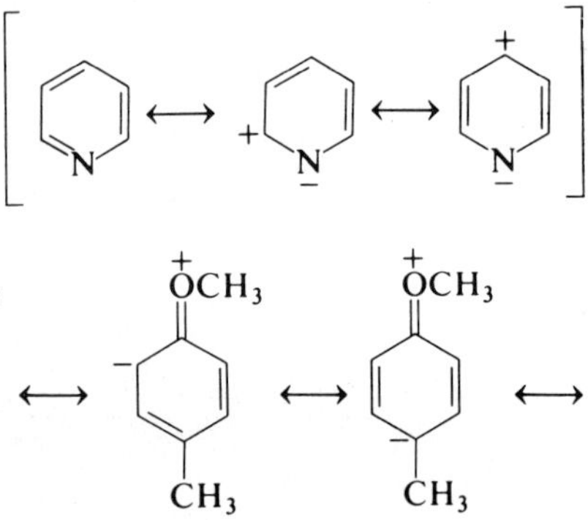

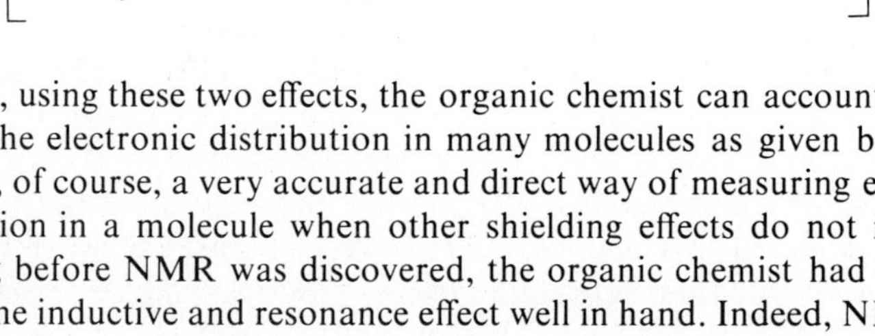

Thus, using these two effects, the organic chemist can account for and predict the electronic distribution in many molecules as given by NMR. NMR is, of course, a very accurate and direct way of measuring electronic distribution in a molecule when other shielding effects do not interfere. But long before NMR was discovered, the organic chemist had the concept of the inductive and resonance effect well in hand. Indeed, NMR only confirmed the electron distribution predicted by these effects. These effects were initially employed in explaining the chemistry of certain organic molecules by accounting for their electron distribution by the inductive and resonance effects. The study of the strengths of acids and bases illustrates this approach, which is introduced in Chapter 3 and discussed in greater detail in R. Stewart's book in this series, *The Investigation of Organic Reactions.*

The *position* of an NMR signal is referred to as its *chemical shift*, since the position of the peak is determined by the electronic environment of the nucleus, which is also responsible for its chemistry. The electron density around the nucleus is the most important factor in determining chemical shifts. In general, the cloud of electrons around one nucleus will not affect the chemical shift of an adjacent nucleus. The reason for this is that the electron clouds around most nuclei are spherical and it can be shown that the resultant magnetic field at nucleus B induced by a group of spherical electrons around an adjacent atom A will be zero when *all* positions of B relative to A and the applied magnetic field are taken into account (see Fig. 2-21). However, if an adjacent group of electrons is not spherically

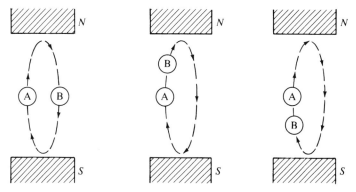

Fig. 2-21 The effect of the induced magnetic field of an adjacent atom at different positions with respect to the magnetic field.

shaped, the resultant magnetic field at a nucleus outside this group of electrons may not be zero. The π-cloud of a benzene ring clearly illustrates this phenomenon. Only when the π-cloud is perpendicular to the applied magnetic field will its electrons be able to circulate. When the electrons do circulate, they always generate a magnetic field at the attached hydrogens which is *in the same direction* as the applied field (see Fig. 2-22). Thus, aromatic hydrogens give rise to a signal much farther downfield than would be expected from consideration of electron density only. The electron density of an aromatic hydrogen atom should be about the same as that of an olefinic hydrogen atom, since both types of carbon-hydrogen bonds involve sp^2 hybrids from the carbon atom. Yet aromatic hydrogen atoms give NMR signals about 2 p.p.m. downfield from the NMR signals of olefinic hydrogen atoms.

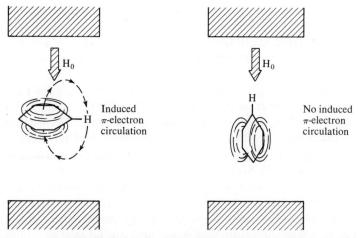

Fig. 2-22 The induced magnetic field of benzene at different positions with respect to the applied magnetic field.

A group of electrons which generates a magnetic field whose magnitude is dependent on the direction of the applied field is said to be *magnetically anisotropic*. Two important groups of atoms that are magnetically anisotropic are aromatic systems and carbonyl groups. The shielding and deshielding areas around these groups are depicted in Fig. 2-23. The closer a nucleus is to the center of one of the regions, the more it will be affected.

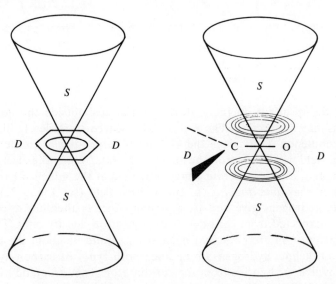

Fig. 2-23 Shielding (S) and deshielding (D) volumes of an aromatic ring and carbonyl group.

2.6 Ultraviolet-Visible, Infrared, and Nuclear Magnetic Resonance Spectroscopy

Because these two factors, electron density and magnetic fields from magnetically anisotropic groups, are relatively constant for a given type of hydrogen no matter what the specific chemical environment is, each type of hydrogen shows up in a certain area of the NMR spectrum. In Table 2-8 are presented the positions of the NMR signals for many common types of hydrogens. There are some exceptions to the values presented in Table 2-8, but these are rare. The hydrogens of hydroxyl groups and amines are not given in Table 2-8, since their positions are highly dependent both on their rate of exchange with other hydroxylic hydrogens and on the amount and chemical shifts of these other hydroxylic hydrogens, and thus vary considerably.

Table 2-8

POSITION OF NMR SIGNALS OF VARIOUS KINDS OF HYDROGEN ATOMS

Type of hydrogen atom	δ-values
C—H (aliphatic)	1.5–0.9
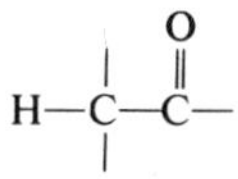 H—C—C(=O)—	2.7–2.0
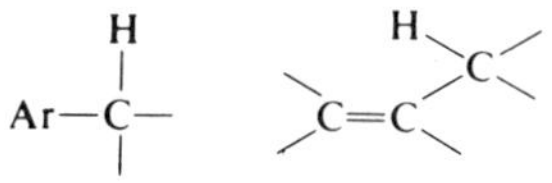 Ar—CH—, C=C—H	3.0–2.2
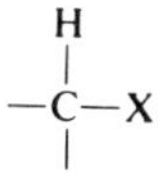 —CH—X (X=Cl, Br, I)	4.0–2.5
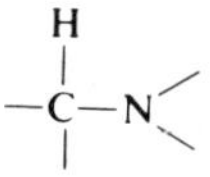 —CH—N<	4.0–3.0
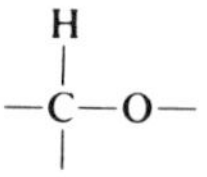 —CH—O—	4.0–3.4
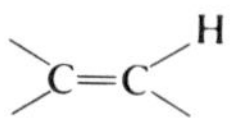 C=C—H	5.9–4.6
C—H (aromatic)	8.5–6.0
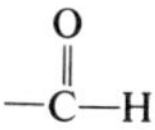 —C(=O)—H	10–9
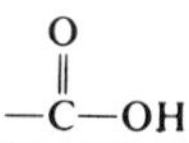 —C(=O)—OH	12–11

In addition to the *chemical shift*, *spin-spin splitting* is another phenomenon of NMR that is information-packed. The NMR spectrum of ethyl chloride given in Fig. 2-20, p. 86, is a poorly resolved one. Higher resolution leads to the spectrum shown in Fig. 2-24, where the low field

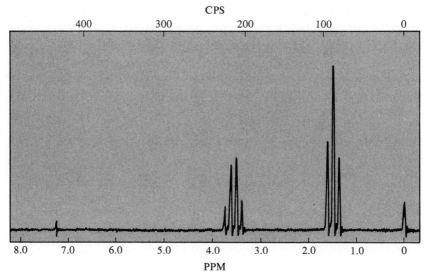

Fig. 2-24 Highly resolved NMR spectrum of ethyl chloride. (From *Varian Spectra Catalog of High Resolution NMR*, Varian Associates, Palo Alto, Calif., 1962, Spectrum 11.)

signal is split into a quartet (intensity 1:3:3:1) and the high field signal is split into a triplet (intensity 1:2:1). In general, any group of equivalent protons on an adjacent atom causes the NMR signal of the hydrogen to be split into $n + 1$ peaks, where n is the number of equivalent protons.

$$H\diagdown_{A-B}\diagup H$$

Thus, the signal of the methylene protons of ethyl chloride is split into a quartet by the methyl group which is, in turn, split into a triplet by the methylene group.

The reason splitting occurs is that an adjacent hydrogen can be in one spin state or the other, say α or β; if it is in one spin state, it will add to the magnetic field of the adjacent proton and if it is in the other spin state, it will subtract from this field. One proton can be either in an α or β state so that it will split the NMR signal of adjacent protons into a doublet with both lines of equal intensity. Two protons can be in spin states $\alpha\alpha$, $\alpha\beta$, $\beta\alpha$, or $\beta\beta$ and will split the NMR signal of adjacent protons into a triplet with the center line twice the intensity of the side lines, since the spin states

$\alpha\beta$ and $\beta\alpha$ affect an adjacent proton equally. In a similar fashion, three protons have spin states $\alpha\alpha\alpha$, $\alpha\alpha\beta$, $\alpha\beta\alpha$, $\beta\alpha\alpha$, $\alpha\beta\beta$, $\beta\alpha\beta$, $\beta\beta\alpha$, and $\beta\beta\beta$, which leads to a quartet in the ratio 1 : 3 : 3 : 1.

The magnitude of the splitting is given by the *coupling constant*, *J*. *J* is independent of the applied magnetic field and is usually given in cycles per second or Hertz (1 Hz = 1 cps). The field independence of *J* is consistent with a coupling mechanism through the bonds. In other words, the spin states of adjacent hydrogens are transmitted through bonds, not space. Unlike *J*, the chemical shift of an atom is dependent on the applied magnetic field since the induced magnetic field, which produces the chemical shift, is a function of the applied field. These two phenomena, the chemical shift and spin-spin splitting, make NMR a powerful tool for structural determination. Not only can one very quickly ascertain the number and kinds of protons in a molecule (by the chemical shift) but one can often also determine which protons are adjacent or not adjacent (by spin-spin splitting). The few simple problems at the end of this chapter illustrate the power of this technique and show that our concepts of molecular structure well explain NMR spectra even though many of the concepts were developed by very indirect reasoning. J. R. Dyer's book in this series, *Applications of Absorption Spectroscopy of Organic Compounds*, covers absorption spectroscopy in much greater detail.

PROBLEMS

1. Match the following compounds with their boiling points (which are listed after the compounds).

$(CH_3)_4C$, $(CH_3)_4Ge$, $(CH_3)_4Pb$, $(CH_3)_4Si$, $(CH_3)_4Sn$;
10°, 27°, 44°, 78°, 110°.

2. Match the following compounds with their boiling points.

$CH_3(CH_2)_6CH_3$, $CH_3(CH_2)_4CH(CH_3)_2$, $(CH_3)_2CHCH_2C(CH_3)_3$;
99°, 116°, 126°.

3. A hydrocarbon with the formula C_8H_{18} has an mp of 104° and a bp of 107°. No other isomer of it is a liquid over such a small temperature range. What is its structure?

4. Which compound of each pair would have the higher melting point and which one would have the higher boiling point?

(a) $(CH_3)_2C=CH_2$ $(CH_3)_2C=O$

(b) cyclopentanone   cyclohexanone

(c) $Ph-NO_2$ $Ph-CH(CH_3)_2$

(d) $CH_3CH_2OCH_2CH_3$ $CH_3CH_2CH_2CH_2OH$

(e) $(CH_3)_3N$ $CH_3—NH—CH_2CH_3$

(f) $CH_3—NH—CH_2CH_3$ $CH_3CH_2CH_2—NH_2$

(g) CH_3COOCH_3 CH_3CH_2COOH

(h) $CH_3(CH_2)_4CH_3$

(i) $C_6H_5—OH$ $C_6H_5—SH$

(j) $HCONHCH_2CH_3$ $HCON(CH_3)_2$

5. Match the following compounds with their melting points.
(a) $o\text{-}C_6H_4Cl_2, p\text{-}C_6H_4Cl_2$; $-18°, 53°$

(b) *cis*-Stilbene, *trans*-stilbene; $-5°, 124°$

(c)

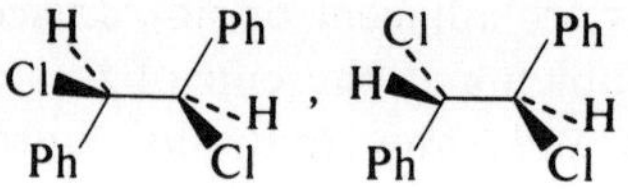

   H   Ph   Cl   Ph
Cl\\\\\\\\C—C////H , H\\\\\\\\C—C////H ; 93°, 192°
 Ph   Cl   Ph   Cl

(d) $1,2,4,5\text{-}(CH_3)_4C_6H_2, 1,2,3,4\text{-}(CH_3)_4C_6H_2$; $-4°, 80°$

(e) *cis*-$HO_2C—CH=CH—CO_2H$ (maleic acid),
trans-$HO_2C—CH=CH—CO_2H$ (fumaric acid); $131°, 287°$

6. Which compound of each pair would have the greater solubility in water? Which one would have the shorter retention time on a nonpolar glpc column?

(a) $p\text{-}CH_3—C_6H_4—SO_3H$ $C_6H_5—SO_3—CH_3$

(b)

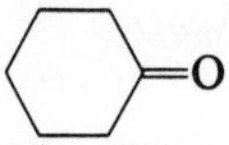

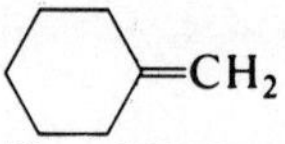

(c) $Ph—NO_2$ $Ph—CH(CH_3)_2$

(d)

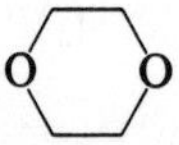

(e) $CH_3(CH_2)_5OH$ $CH_3(CH_2)_3CHOHCH_3$

7. Which compound of each pair would have the higher λ_{max}?

(a)

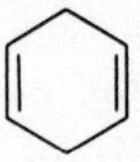

(b) $Ph—Ph$ $Ph—CH_2—Ph$

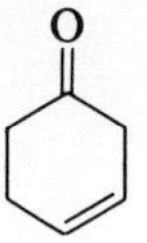

(c)

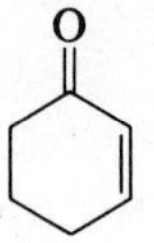

(d)

(e) C₆H₅—NO₂   C₆H₅—NH₂
(f) C₆H₅—CH₂COOH   C₆H₅COOH

8. Draw the p-orbitals that make up the π-orbitals of the planar form of 1,3-butadiene. Use lines to represent all σ-bonds.

9. Indicate which rule or principle if any is being broken by electronic configurations for ground states shown on page 98.

10. Draw reasonable resonance structures for the following two ions. Which ion should be more stable?

11. How many magnetically different hydrogen atoms (i.e., hydrogen atoms with different NMR chemical shifts) do the compounds in Problem 7 have?

12. Give the structures for the following compounds based on their molecular formula, major infrared absorptions, and NMR data. The NMR data is reported in ppm from tetramethylsilane and s = singlet, d = 1:1 doublet, t = 1:2:1 triplet, q = 1:3:3:1 quartet, and m = multiplet.

(a) $C_3H_6Cl_2$; IR: 2960 cm⁻¹; NMR: δ 3.70 (t, 4H, J = 6 Hz), 2.20 (quintet, 2H, J = 6 Hz).

(b) $C_3H_3Cl_5$; IR: 2960 cm⁻¹; NMR: δ 6.1 (d, 2H, J = 6 Hz), 4.2 (t, 1H, J = 6 Hz).

(c) $C_4H_8O_2$; IR: 2950, 1100 cm⁻¹; NMR: δ 3.70 (s, 8H).

(d) $C_4H_8O_3$; IR: 3400–2800 (broad), 1690, 1100 cm⁻¹; NMR: δ 10.95 (s, 1H), 4.13 (s, 2H), 3.66 (q, 2H, J = 7 Hz), 1.27 (t, 3H, J = 7 Hz).

(e) $C_5H_{10}O$; IR: 2980, 1720 cm⁻¹; NMR: δ 2.6 (septet, 1H, J = 7 Hz), 2.1 (s, 3H), 1.0 (d, 6H, J = 7 Hz).

(f) $C_6H_{12}O_2$; IR: 2960, 1735, 1100 cm⁻¹; NMR: δ 2.1 (s, 3H), 1.2 (s, 9H).

(g) $C_6H_{10}O$; IR: 2950, 1780 cm⁻¹; NMR: δ 2.1 (s, 4H), 1.1 (s, 6H).

(h) $C_7H_{12}O_4$; IR: 2950, 1735, 1100 cm⁻¹; NMR: δ 4.1 (q, 4H, J = 7 Hz), 3.3 (s, 2H), 1.1 (t, 6H, J = 7 Hz).

(i) $C_7H_{12}O_3$; IR: 2950, 1735, 1710, 1100 cm⁻¹; NMR: δ 3.8 (s, 3H), 3.1 (s, 2H), 2.1 (septet, 1H, J = 7 Hz), 1.1 (d, 6H, J = 7 Hz).

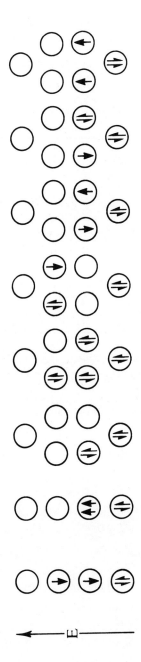

(j) $C_8H_{10}O$; IR: 3050, 2910, 1100 cm^{-1}, NMR: δ 7.3 (sharp m, 5H), 4.3 (s, 2H), 3.4 (s, 3H).

(k) $C_8H_{10}O$; IR: 3500, 3050, 2950, 1100 cm^{-1}; NMR: δ 7.1 (sharp m, 5H), 4.6 (q, 1H, J = 7 Hz), 2.9 (s, 1H), 1.8 (d, 3H, J = 7 Hz).

(l) $C_8H_{14}O_4$; IR: 2960, 1735, 1100 cm^{-1}; NMR: δ 3.9 (q, 4H, J = 7 Hz), 2.4 (s, 4H), 1.1 (t, 6H, J = 7 Hz).

(m) $C_8H_{14}O_2$; IR: 2950, 1735, 1100 cm^{-1}; NMR: δ 3.9 (d, 2H, J = 7 Hz), 2.1 (s, 3H), 1.6–1.1 (m, 9H).

(n) C_8H_{14}; IR: 3050, 2960; NMR: δ 6.0 (s, 2H), 1.8 (s, 6H), 1.7 (s, 6H).

(o) $C_9H_6O_6$; IR: 3200–2500 (broad), 1700, 1100 cm^{-1}; NMR: δ 12.0 (s, 3H), 8.0 (s, 3H).

(p) $C_9H_{11}NO$; IR: 3050, 2950, 1690 cm^{-1}; NMR: δ 7.6 (sharp m, 5H), 2.6 (s, 6H).

(q) $C_9H_7NO_2$; IR: 3030, 2960, 2200, 1735, 1100 cm^{-1}; NMR: δ 7.6 (A_2B_2 quartet that is characteristic of *para*-disubstituted benzene rings with the substituents being different, 4H), 4.3 (s, 3H).

(r) $C_{10}H_{13}Cl$; IR: 3050, 2960 cm^{-1}; NMR: δ 7.2 (sharp m, 5, H), 3.0 (s, 2H), 1.3 (s, 6H).

(s) $C_{10}H_{12}O$; IR: 3010, 2960, 1710 cm^{-1}; NMR: δ 10.4 (s, 1H), 6.9 (s, 2H), 2.8 (s, 6H), 2.2 (s, 3H).

(t) $C_{10}H_{10}O_4$; IR: 1735, 1725, 1100 cm^{-1}; NMR: δ 7.3 (A_2B_2 quartet, 4H), 4.0 (s, 3H), 2.1 (s, 3H).

(u) $C_{10}H_{12}O_2$; IR: 3050, 2960, 1735, 1100 cm^{-1}, NMR: δ 7.3 (sharp m, 5H), 4.3 (t, 2H, J = 7 Hz), 2.8 (t, 2H, J = 7 Hz), 2.1 (s, 3H).

(v) $C_{10}H_{14}O_2$; IR: 3500, 1100; NMR: δ 7.1 (s, 4H), 4.8 (q, 2H, J = 7 Hz); 3.8 (s, 2H), 1.3 (d, 6H, J = 7 Hz).

(w) $C_{10}H_{12}O_3$, IR: 3500, 3050, 2950, 1720, 1100 cm^{-1}; NMR: δ 7.5 (A_2B_2 quartet, 4H), 6.0 (s, 1H), 4.6 (s, 3H), 4.2 (t, 2H, J = 7 Hz), 2.0 (t, 2H, J = 7 Hz).

(x) $C_{15}H_{12}O$; IR: 3050, 2900, 1710 cm^{-1}; NMR: δ7.6–7.2 (m, 8H), 3.6 (s, 4H).

(y) $C_{15}H_{14}O$; IR: 3050, 2950, 1710 cm^{-1}; NMR: δ 7.2 (sharp m, 10H), 5.1 (s, 1H), 2.2 (s, 3H).

(z) $C_{21}H_{18}O_2$; IR: 3010, 2960, 1735, 1100 cm^{-1}; NMR: δ 7.1 (sharp m, 15H), 5.3 (s, 2H), 5.1 (s, 1H).

REFERENCES

1. L. C. Pauling and R. Haywood, *The Architecture of Molecules*. San Francisco: W. H. Freeman, 1964.

2. Jack Hine, *Physical Organic Chemistry*, 2nd ed. New York: McGraw-Hill, 1962, pp. 3–9.

3. D. J. Pasto and C. R. Johnson, *Organic Structure Determination*. Englewood Cliffs, N.J.: Prentice-Hall, 1969.

4. J. B. Hendrickson, D. J. Cram, and G. S. Hammond, *Organic Chemistry*, 3rd Ed. New York: McGraw Hill, 1970, Ch. 7.

5. R. M. Silverstein and G. C. Bassler, *Spectrometric Identification of Organic Compounds*, 2nd ed. New York: Wiley, 1967.

6. L. M. Jackman and S. Sternhell, *Applications of Nuclear Magnetic Resonance Spectroscopy in Organic Chemistry*, 2nd ed. London: Pergamon, 1969.

7. K. Nakanishi, *Infrared Absorption Spectroscopy*. San Francisco; Holden-Day, 1962.

3
Chemical Interrelations of Functional Groups

3.1 REACTIONS OF FUNCTIONAL GROUPS

When 1-pentene is treated with bromine in carbon tetrachloride, the red-brown bromine color rapidly disappears. This same bromine decolorization occurs if 1-hexene, *cis*-or *trans*-2-hexene, cyclohexene, or almost any other alkene is used in place of 1-pentene. In fact, a characteristic test for the presence of an alkene or olefin is the decolorization of bromine. Obviously the bromine is reacting with the carbon-carbon double bond in a fashion that is virtually independent of the specific molecular environment of the olefin. *That a functional group, such as a carbon-carbon double bond, undergoes a characteristic reaction that is largely independent of the molecular environment of the functional group is the situation that is generally found in organic chemistry.* Although the specific molecular environment of a functional group can affect its reactivity in minor or major ways and organic chemists spend much time studying these variations, nevertheless, for a given functional group, a set of reactions exists that this functional group will generally undergo. Just as functional groups lead to certain physical properties, they lead to certain chemical properties. The organic chemist can, therefore, synthesize new molecules with certain desired physical and chemical properties since he can predict the reactions of functional groups both during the synthesis and in the new molecule.

In this chapter, various classes of reactions will be presented that convert one functional group to another. Classes of reactions are presented, since the student will find that reactions are easier to remember if they are grouped in classes. Moreover, relationships between various classes of reactions make mechanistic similarities between the classes of reactions clearer. Finally, consideration of reactions of functional groups as classes often makes the chemical similarities of various functional groups more vivid.

Although in the following discussion reactions are presented in classes as reaction types, the student should be aware throughout of what functional groups are undergoing the particular reactions.

3.2 REACTIONS OF ALKENES (OLEFINS) AND ALKYNES (ACETYLENES)—ADDITION REACTIONS

Let us return to the reaction of bromine with alkenes. Examination of the products of the reaction shows that the bromine *adds* to the alkene to form a 1,2-dibromide. As would be expected from our discussion of

$$\diagup\!\!=\!\!\diagdown + Br\!-\!Br \longrightarrow \diagdown\!\!-\!\!\diagup$$
$$ Br \quad Br$$
$$\text{red-brown} \text{colorless}$$

visible absorption spectroscopy, the dibromide is colorless. This type of reaction is, quite reasonably, known as an *addition reaction*. Not only does bromine add to an alkene, but many molecules do. For example, the hydrogen halides, water, hydrogen, and borane all add to alkenes. *Addition reactions* constitute the first major class of reactions that we will discuss. It is obvious how useful the first four of these reactions are since they permit the conversion of alkenes to alkyl halides, alcohols, alkanes, and 1,2-alkanediols. The last reaction, the addition of borane, is important because the resultant alkyl borane can be oxidized to an alcohol with hydrogen peroxide. Why this sequence of reactions is often used to con-

$$\diagup\!\!=\!\!\diagdown + H\!-\!X \longrightarrow \diagdown\!\!-\!\!\diagup \quad (X = Cl, Br, I)$$
$$ H \quad X$$

$$\diagup\!\!=\!\!\diagdown + H\!-\!OH \xrightarrow[\text{(catalyst)}]{H^+} \diagdown\!\!-\!\!\diagup$$
$$ H \quad OH$$

$$\diagup\!\!=\!\!\diagdown + H\!-\!H \xrightarrow[\text{(catalyst)}]{Pt} \diagdown\!\!-\!\!\diagup$$
$$ H \quad H$$

$$\diagup\!\!=\!\!\diagdown + OsO_4 \text{ (or } MnO_4^-) \longrightarrow \diagdown\!\!-\!\!\diagup$$
$$ OH \quad OH$$

$$\diagup\!\!=\!\!\diagdown + H\!-\!BH_2 \longrightarrow \diagdown\!\!-\!\!\diagup$$
$$ H \quad BH_2$$

3.2 Reactions of Alkenes and Alkynes—Addition Reactions

vert an alkene into an alcohol instead of acid-catalyzed hydration of an alkene will be explained on p. 104.

$$\underset{H\quad BH_2}{\diagup\!\!\!\diagdown\!\!\!\!\diagdown\!\!\!\diagup} \xrightarrow{H_2O_2} \underset{H\quad OH}{\diagup\!\!\!\diagdown\!\!\!\!\diagdown\!\!\!\diagup}$$

If the two sides of a carbon-carbon double bond are different, two different products are possible when an unsymmetrical molecule is added to the alkene. For example, hydrochloric acid could add to 1-hexene to give 1- or 2-chlorohexane. Many years ago (1869) Markovnikov observed that if H-Z was added to an alkene that contained a different number of hydrogens on each carbon,

$$CH_3CH_2CH_2CH_2CH=CH_2 \xrightarrow{H-Cl} CH_3CH_2CH_2CH_2CH_2CH_2-Cl$$

or

$$CH_3CH_2CH_2CH_2\underset{Cl}{\overset{|}{C}HCH_3}$$

the H of H-Z went to the carbon that already had the most hydrogens ("those that have, get more"). Thus hydrogen chloride reacts with 1-hexene to give mainly 2-chlorohexane. This rule, now known as *Markovnikov's rule*, was an empirical rule, since mechanisms were poorly understood in 1869. Today, we know that these additions can be looked upon as going in two steps, with the intermediate formation of a carbonium ion. Since tertiary carbonium ions are more stable than secondary carbonium ions, which in turn are more stable than primary carbonium ions,* the more stable ion will form preferentially if protonation of an

$$CH_3(CH_2)_3-CH=CH_2 + H^+ \longrightarrow CH_3(CH_2)_3\overset{+}{C}H-CH_3$$
2° carbonium ion

$$CH_3(CH_2)_3-\overset{+}{C}H-CH_3 + Cl^- \longrightarrow CH_3(CH_2)_3\underset{Cl}{\overset{|}{C}H}-CH_3$$

unsymmetrical alkene leads to two different ions. In the case of 1-hexene, the 2-cation forms, rather than the 1-cation, since the former is secondary

$$CH_3(CH_2)_3-CH=CH_2 + H^+ \longrightarrow CH_3(CH_2)_3-CH_2-CH_2^+$$
1° carbonium ion

and the latter is primary. These mechanistic considerations beautifully account for Markovnikov's empirical rule.

Reagents that attack alkenes to produce cations are called *electrophiles*

---

* N. L. Allinger and J. Allinger, *Structures of Organic Molecules*, in this series, pp. 111–113.

since they are "electron seeking." Such reactions are called *electrophilic reactions* and are by far the most common type of reaction that alkenes undergo.

Many times it would be useful to be able to bring about "anti-Markovnikov addition," or addition of H-Z in the opposite direction to that predicted by Markovnikov's rule. A chemist might very well want a 1-isomer, not a 2-isomer. In the case of the addition of water (which leads

$$CH_3(CH_2)_3CH=CH_2 + H\text{-}Z \xrightarrow{?} CH_3(CH_2)_3\underset{H}{\overset{|}{C}}H-\underset{Z}{\overset{|}{C}}H_2$$

to an alcohol) H. C. Brown has worked out a very clever sequence of reactions which leads to "anti-Markovnikov addition." The sequence of reactions centers around the fact that the boron atom of borane, BH_3, is electron deficient. The reaction of borane* with alkenes can be thought of as leading to the formation of a cation which rapidly picks up a hydride ion (H^-) from the boron. Now, considering this mechanism, one predicts that the addition of borane to 1-hexene should give the 1-isomer, and

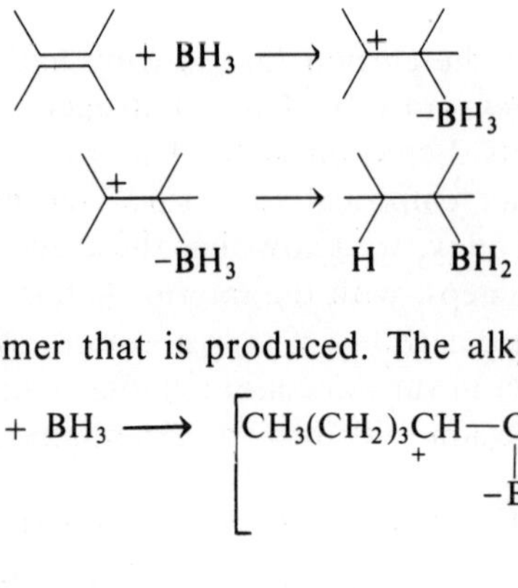

indeed this is the isomer that is produced. The alkylborane can be con-

$$CH_3(CH_2)_3CH=CH_2 + BH_3 \longrightarrow \left[CH_3(CH_2)_3\underset{-BH_3}{\overset{+}{C}H-CH_2} \right] \longrightarrow$$

$$CH_3(CH_2)_3\underset{H}{\overset{|}{C}}H-\underset{BH_2}{\overset{|}{C}}H_2$$

veniently converted to the alcohol by oxidation with hydrogen peroxide and overall "anti-Markovnikov" addition of water is achieved.

$$CH_3(CH_2)_3CH_2-CH_2-B\hspace{-0.5em}\diagdown + H_2O_2 \longrightarrow$$
$$CH_3(CH_2)_3CH_2-CH_2-OH + HO-B\hspace{-0.5em}\diagdown$$

Some addition reactions form rings, since both ends of the carbon-carbon double bond become attached to the same atom or group of atoms. Epoxidation of alkenes and ozonolysis are two such reactions.

---

* Borane, BH_3, is actually dimerized, $(BH_3)_2$, but many of its reactions can be described in terms of the monomer.

3.2 Reactions of Alkenes and Alkynes—Addition Reactions

Epoxides are prepared by treating an alkene with a peracid (or peroxy-

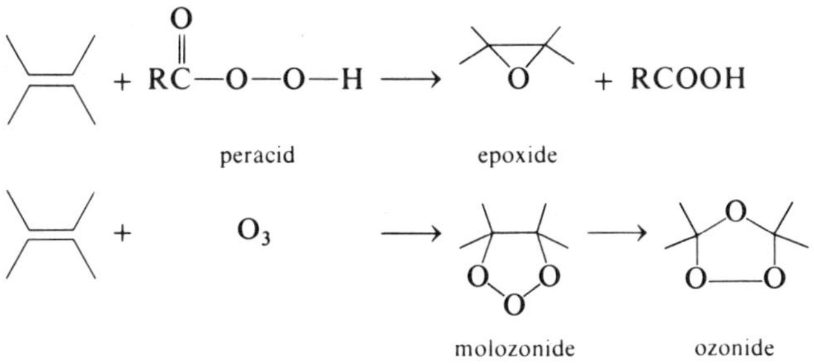

$$\text{alkene} + RC(=O)-O-O-H \longrightarrow \text{epoxide} + RCOOH$$

peracid → epoxide

$$\text{alkene} + O_3 \longrightarrow \text{molozonide} \longrightarrow \text{ozonide}$$

acid) and are stable and useful products themselves. Reaction of alkenes with ozone initially gives molozonides, which are very reactive and rearrange rapidly to ozonides. Although ozonides are relatively stable

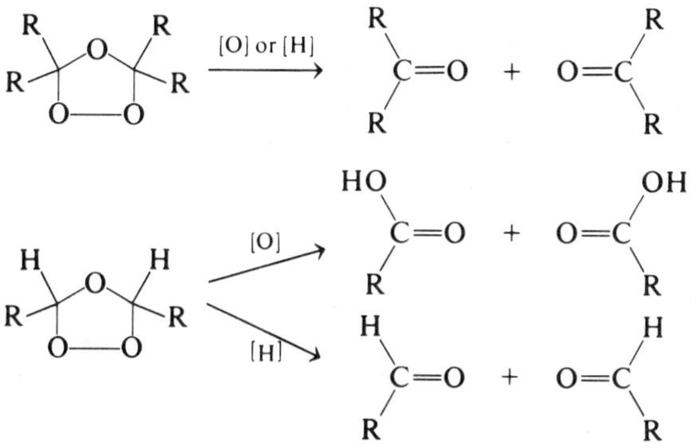

$$\text{ozonide} \xrightarrow{[O] \text{ or } [H]} R_2C=O \;+\; O=CR_2$$

$$\text{ozonide (H,R)} \xrightarrow{[O]} \substack{HO\\R}C=O \;+\; O=C\substack{OH\\R}$$

$$\xrightarrow{[H]} \substack{H\\R}C=O \;+\; O=C\substack{H\\R}$$

compounds, normally they are oxidized to ketones and acids or reduced to aldehydes and ketones. The overall reaction, then, splits the alkene into two parts. Since often the aldehydes, ketones, or acids are compounds with known structures, the structure of an alkene can be determined by treating the alkene with ozone followed by an oxidative or reductive workup and determination of the products of the reactions. For example, if an alkene of unknown structure gives butanal and 2-butanone after treatment with ozone followed by a reductive workup, the alkene

$$\text{Olefin} \xrightarrow[(2)\,[H]]{(1)\,O_3} CH_3CH_2CH_2CH=O \quad CH_3\overset{O}{\underset{\|}{C}}CH_2CH_3$$

must be 3-methyl-3-heptene,

$$CH_3CH_2CH_2CH=C\genfrac{}{}{0pt}{}{CH_3}{CH_2CH_3}$$

The addition of carbenes and carbenoid species to alkenes to form cyclopropanes is a very important reaction. For example, ethyl diazoacetate can be decomposed photochemically or thermally to give the expected carbene, which will form a cyclopropane by addition to a double bond.

$$\text{>=<} + N_2CH-COOEt \xrightarrow[\text{or } \Delta]{h\nu} [:CH-COOEt] + N_2 \longrightarrow \text{cyclopropane with H, COOEt}$$

The reaction of an alkene with methylene diiodide in the presence of a zinc-copper couple is a very good method of preparing cyclopropanes.

$$\text{>=<} + CH_2I_2 \xrightarrow[\text{couple}]{Zn-Cu} \text{cyclopropane-}CH_2$$

The reaction is considered to be a "carbenoid" reaction since, although the reaction looks like the addition of a carbene to an alkene, no free carbene can be detected and side reactions characteristic of carbenes are absent. This reaction was discovered and developed by Simmons and Smith and is consequently called the *Simmons-Smith reaction*. It is quite common for organic chemists to refer to a general reaction by a chemist's or chemists' names and this is one of many name reactions the student should remember. Although this often seems like an unnecessary burden on a student's memory, it is not, since it is easier to refer to reactions by chemists' names than by a short description of the reactions. Mention of a name reaction conveys reactants, products, conditions, mechanism, and many other things to an informed chemist.

The additions so far discussed have been *1,2-additions* since, indeed, the reagents have been added to adjacent atoms. When addition occurs to a conjugated diene system, 1,4- as well as 1,2-addition takes place. Often one or the other type of addition will predominate, but the reaction conditions can drastically affect the ratio of the two types of additions.

$$\begin{array}{c} CH_2 \\ \| \\ CH \\ | \\ CH \\ \| \\ CH_2 \end{array} + HCl \longrightarrow \underset{\text{1,2-addition}}{CH_2-CH-CH=CH_2} \quad \underset{\text{1,4-addition}}{CH_2-CH=CH-CH_2}$$
with H, Cl on first two carbons (1,2) and H, Cl on terminal carbons (1,4)

An extremely important reaction which always involves 1,4-addition is the Diels-Alder reaction. This is the addition of an alkene, called a *dienophile*, to a conjugated diene to form a six-membered ring. The

3.2 Reactions of Alkenes and Alkynes—Addition Reactions

reaction takes place thermally and often requires an "activated" alkene for a dienophile. An activated carbon-carbon double bond is one that is attached to some functional group such as a carboalkoxy, acetate, or nitro group. The general name for reactions such as the Diels-Alder reaction is *cycloaddition reactions*. The most common type of thermal cycloaddition reaction is the 4 + 2 reaction (i.e., the reaction between a four-atom π-system and a two-atom π-system). However, the most common type of photochemical cycloaddition reaction is the 2 + 2 reaction. For example, cinnamic acid dimerizes when exposed to light to form various cyclo-

$$C_6H_5-CH=CH-COOH \xrightarrow{h\nu}$$

butane derivatives. Woodward and Hoffmann[1] have recently devised a set of selection rules based on symmetries of molecular orbitals which explain why thermal cycloadditions are 4 + 2 and photochemical cycloadditions are 2 + 2. Reactions such as these and the Woodward-Hoffmann selection rules are discussed in C. H. DePuy and O. L. Chapman's book in this series, *Molecular Reactions and Photochemistry*.

There are many other addition reactions that are important but even the present brief survey of addition reactions should establish their significance. W. H. Saunders' book in this series, *Ionic Aliphatic Reactions*, and K. L. Rinehart's book in this series, *Oxidation and Reduction of Organic Compounds*, discuss nonredox and redox addition reactions, respectively.

As with alkenes, addition reactions to alkynes occur. However, with alkynes, or acetylenes, several new questions are raised. Can one stop at the alkene formed, or will addition to the alkene take place under the reaction conditions? If one can stop at the alkene, will the *cis*- or *trans*-alkene be obtained?

$$-C\equiv C- + HZ \longrightarrow -\underset{H}{\overset{}{C}}=\underset{Z}{\overset{}{C}}-$$

The answer to the first question is that one can stop at the alkene stage in most cases if the reaction conditions are properly chosen. For example, catalytic hydrogenation of an alkyne will lead to an alkane unless a special additive (often called a "poison") is added to decrease the effectiveness of the catalyst. Use of a poisoned catalyst permits formation of an alkene.

$$R-C\equiv CH + 2H_2 \xrightarrow[\text{catalyst}]{\text{normal}} R-CH_2CH_3$$

$$R-C\equiv CH + H_2 \xrightarrow[\text{catalyst}]{\text{poisoned}} R-CH=CH_2$$

If a disubstituted carbon-carbon triple bond is submitted to an addition reaction, normally only one geometric isomer of the alkene will predominate. In the case of hydrogenation, the *cis* isomer is usually exclusively formed. In fact, this is a very good way to prepare the *cis* isomer of an alkene, which is usually the thermodynamically less stable isomer.

$$CH_3-C\equiv C-CH_3 \xrightarrow[\text{catalyst}]{H_2,\text{ poisoned}} \begin{array}{c} CH_3 \\ \diagup \\ H \end{array} C=C \begin{array}{c} CH_3 \\ \diagdown \\ H \end{array}$$

Addition reactions of halogens and hydrogen halides to alkynes occur and can usually be stopped at the alkene stage if desired. In the case of hydrogen halides, Markovnikov's rule applies.

$$CH_3-C\equiv CH \xrightarrow{HCl} \begin{array}{c} CH_3 \\ \diagup \\ Cl \end{array} C=CH_2 \xrightarrow{HCl} CH_3-\underset{\underset{Cl}{|}}{\overset{\overset{Cl}{|}}{C}}-CH_3$$

3.3 PREPARATION OF ALKENES (OLEFINS) AND ALKYNES (ACETYLENES)—ELIMINATION REACTIONS

Another major class of reactions consists of *elimination reactions*, which are simply the reverse of addition reactions. Elimination reactions are obviously a general way of forming alkenes and alkynes. All elimination reactions involve the loss of an atom or group of atoms, E^+, which

$$-\underset{\underset{E}{|}}{\overset{|}{C}}-\underset{\underset{Z}{|}}{\overset{|}{C}}- \longrightarrow\ \ \diagdown C=C\diagup\ \ + E^+ + Z^-$$

donates a pair of electrons to the molecule, and an atom or group of atoms, Z^-, which removes a pair of electrons from the molecule. The E^+ is often H^+ and corresponds to the electrophile in addition reactions. The atom or group of atoms that is negative, Z^-, is referred to as a *leaving group*. Good *leaving groups* are obviously ones that can bear the additional pair of electrons well. For example, the chloride, bromide, and iodide ions are good leaving groups since they are relatively stable anions. Although leaving groups are often neutral when attached to the molecule and become negatively charged upon leaving the molecule, many are initially positive and become neutral upon leaving.

Three common types of elimination reactions are dehydrohalogenations [Eq. (3.1)], dehydrations [Eq. (3.2)], and dehalogenations [Eq. (3.3)].

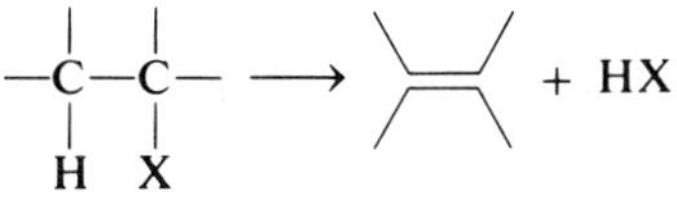

3.3 Preparation of Alkenes and Alkynes—Elimination Reactions

$$-\underset{H}{\overset{|}{C}}-\underset{X}{\overset{|}{C}}- \longrightarrow \;\;\rangle=\langle\;\; + \;HX \quad\quad (3.1)$$

X = Cl, Br, I

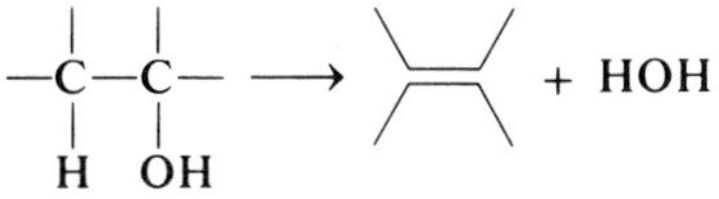

$$-\underset{H}{\overset{|}{C}}-\underset{OH}{\overset{|}{C}}- \longrightarrow \;\;\rangle=\langle\;\; + \;HOH \quad\quad (3.2)$$

$$-\underset{X}{\overset{|}{C}}-\underset{X}{\overset{|}{C}}- \longrightarrow \;\;\rangle=\langle\;\; + \;X_2 \quad\quad (3.3)$$

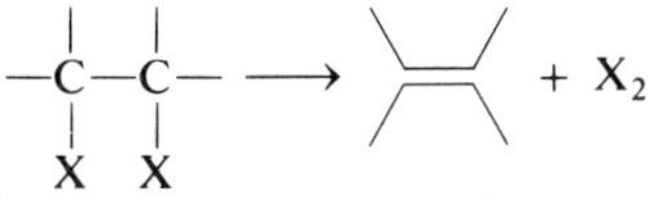

X = Cl, Br, I

Dehydrohalogenations are usually brought about by treating an alkyl halide with base, whereas

[cyclohexyl-Cl] $\xrightarrow{K^+{}^-OH}$ [cyclohexene] + K^+Cl^- + H_2O

dehydrations are usually effected with acid. A common method of

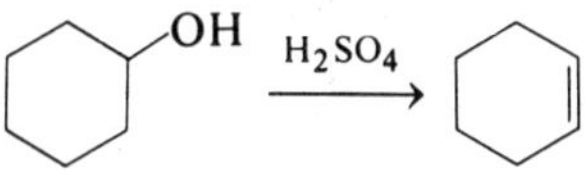

[cyclohexyl-OH] $\xrightarrow{H_2SO_4}$ [cyclohexene]

dehalogenation involves the use of metallic zinc. In dehydrohalogenations,

[cyclohexyl-Br,Br] + Zn $\longrightarrow$ [cyclohexene] + $ZnBr_2$

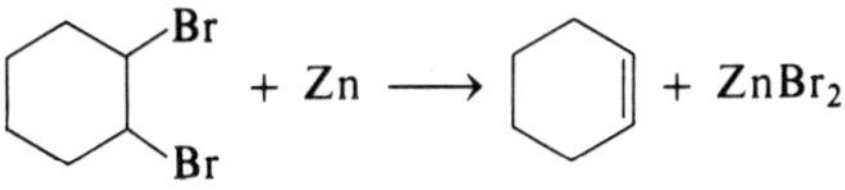

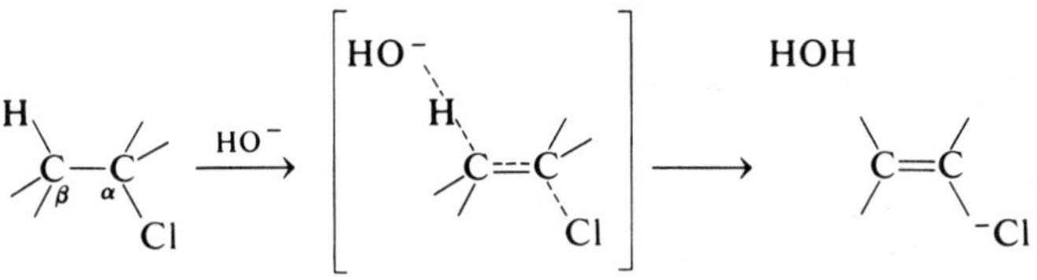

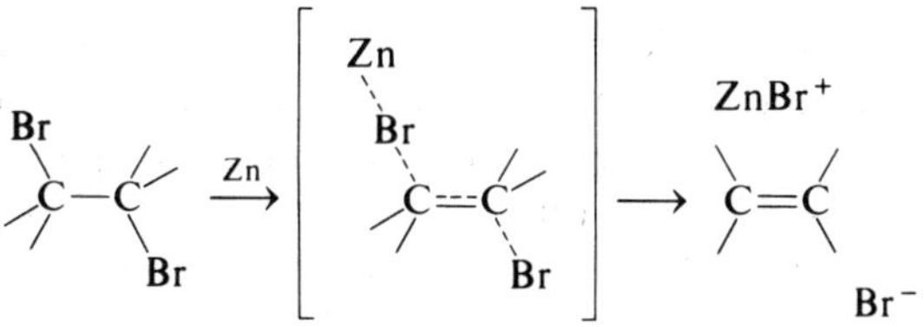

the base, ^-OH, assists the removal of the β hydrogen and in dehalogenations the zinc assists the removal of the β halogen. However, in the dehydration of an alcohol, the acid converts the ⁻OH, a poor leaving group, into OH$_2$, a good leaving group, since it forms the stable molecule

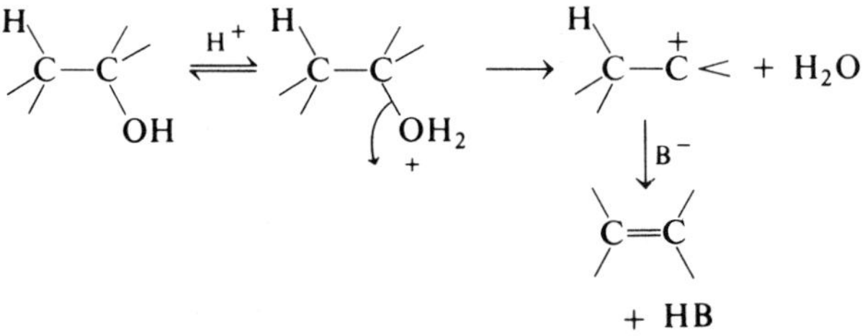

water. The dehydration involves the formation of an intermediate carbonium ion which then loses a β proton to some base, B⁻.

Other types of elimination reactions are known. For example, the elimination of ammonium ions is an important reaction. Positively charged ammonium groups are good leaving groups leading to neutral amines, which are quite stable molecules.

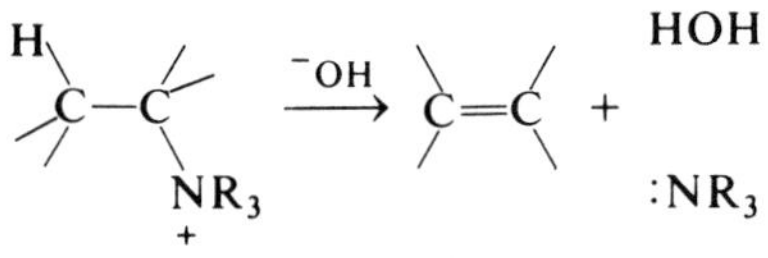

Pyrolysis reactions of some compounds are important elimination reactions. In these cases, the positive and negative groups usually leave together to form a molecule. Pyrolysis of alkyl acetates to form an alkene and acetic acid illustrates this type of elimination reaction. Pyrolysis of amine oxides to form an alkene and a hydroxylamine is another important elimination reaction.

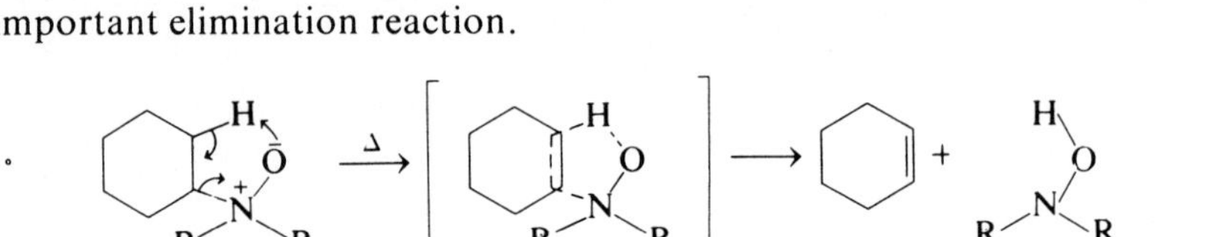

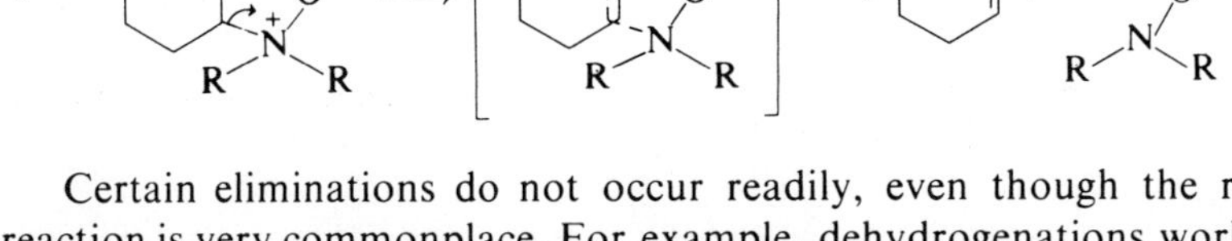

Certain eliminations do not occur readily, even though the reverse reaction is very commonplace. For example, dehydrogenations work only in special cases, even though hydrogenations are quite common and general. Usually dehydrogenations work well only when an aromatic system is being formed. These reactions are more fully discussed in K. L.

3.4 Aromatic Systems—Substitution Reactions

Rinehart's book in this series, *Oxidation and Reduction of Organic Compounds*.

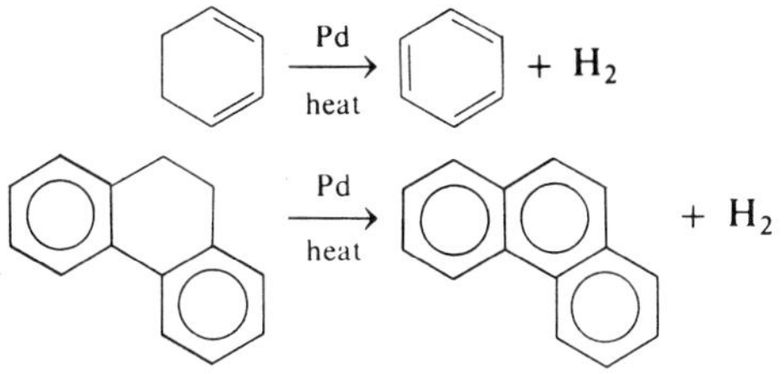

$$\text{cyclohexene} \xrightarrow[\text{heat}]{\text{Pd}} \text{benzene} + H_2$$

$$\text{dihydrophenanthrene} \xrightarrow[\text{heat}]{\text{Pd}} \text{naphthalene derivative} + H_2$$

Alkynes are readily formed by double elimination reactions. Double dehydrohalogenation reactions are quite common. The addition of bromine followed by double dehydrobromination is a convenient method of converting an alkene into an alkyne. Even in systems that can lead to

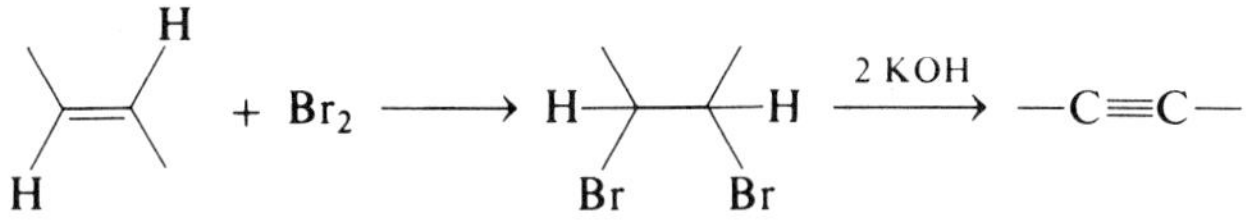

$$\text{CH}_2=\text{CH}_2 + Br_2 \longrightarrow \text{H-CBr-CBr-H} \xrightarrow{2 \text{ KOH}} -C\equiv C-$$

allenes ($-\text{CHBr}-\text{CHBr}-\text{CH}_2- \rightarrow -\text{CH}=\text{C}=\text{CH}-$), the carbon-carbon triple bond is usually formed.

Elimination reactions are more fully discussed in W. H. Saunders' book in this series, *Ionic Aliphatic Reactions*.

3.4 AROMATIC SYSTEMS—SUBSTITUTION REACTIONS

Although addition of bromine to alkenes occurs very rapidly, benzene is essentially inert to bromine. In fact, bromine and benzene can remain

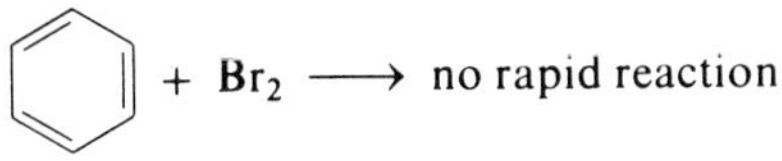

$$\text{benzene} + Br_2 \longrightarrow \text{no rapid reaction}$$

together for days with no reaction. The large amount of unsaturation of benzene makes this a surprising experimental result. In the presence of a suitable catalyst, such as ferric bromide, benzene does react with bromine at moderate temperatures. However, a *substitution reaction* takes place instead of an addition reaction. Substitution reactions constitute a third

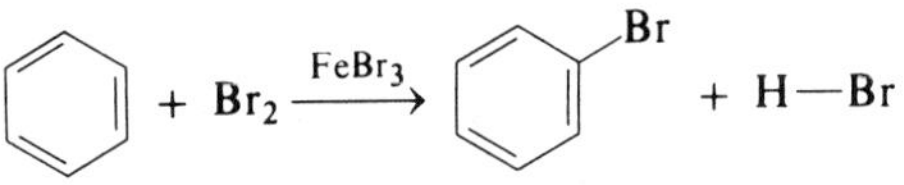

$$\text{benzene} + Br_2 \xrightarrow{\text{FeBr}_3} \text{bromobenzene} + H-Br$$

major class of reactions and include those reactions which replace one atom with another. The mechanism of aromatic substitution is discussed in L. Stock's book in this series, *Aromatic Substitution Reactions*. However, it is obvious that the large stability of an aromatic system* is a driving force for substitution reactions, since substitution reactions do not destroy the aromatic system as addition reactions would.

Several other functional groups can be substituted for hydrogens on an aromatic system. The following reactions represent good examples.

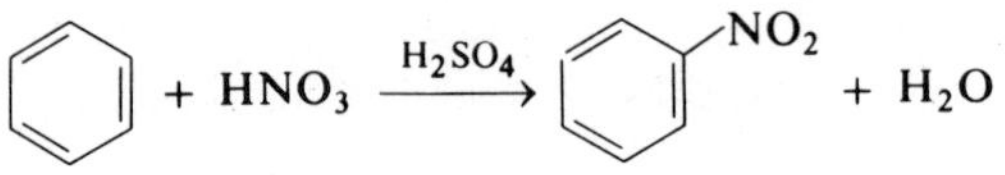

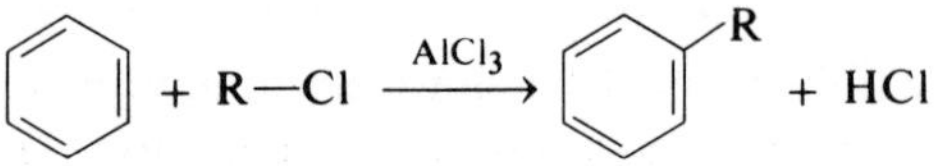

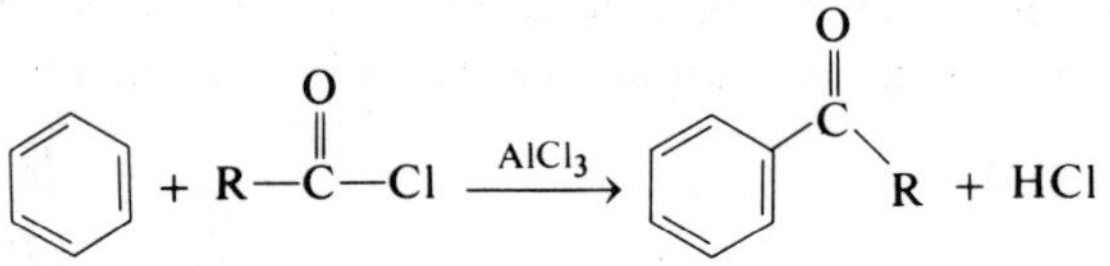

$$C_6H_6 + HNO_3 \xrightarrow{H_2SO_4} C_6H_5NO_2 + H_2O$$

$$C_6H_6 + R-Cl \xrightarrow{AlCl_3} C_6H_5-R + HCl$$

$$C_6H_6 + R-\overset{O}{\underset{\|}{C}}-Cl \xrightarrow{AlCl_3} C_6H_5-\overset{O}{\underset{\|}{C}}-R + HCl$$

The last two reactions, alkylations and acylations of aromatic systems, were discovered and developed by Friedel and Crafts and are consequently called *Friedel-Crafts reactions*.

All aromatic substitution reactions do not involve the replacement of hydrogen. An important reaction is the conversion of an aromatic amine to a diazonium salt, which in turn can be converted to many more stable derivatives.

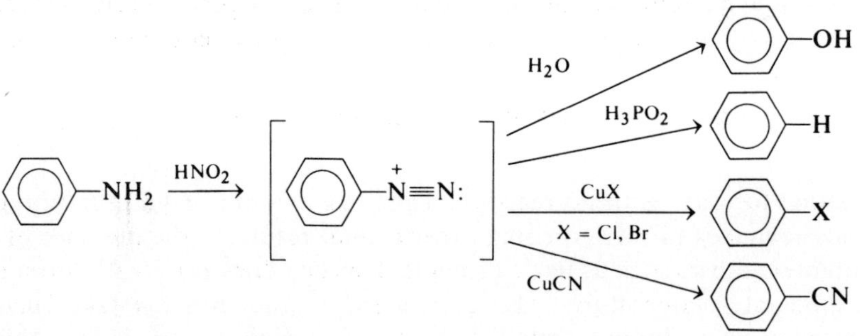

$$C_6H_5-NH_2 \xrightarrow{HNO_2} [C_6H_5-\overset{+}{N}\equiv N:] \begin{array}{l} \xrightarrow{H_2O} C_6H_5-OH \\ \xrightarrow{H_3PO_2} C_6H_5-H \\ \xrightarrow{CuX} C_6H_5-X \quad X = Cl, Br \\ \xrightarrow{CuCN} C_6H_5-CN \end{array}$$

Another extremely important reaction which involves the conversion of a halide to many other groups is the *Grignard reaction*. The first step of the Grignard reaction is the preparation of the *Grignard reagent*. The Grignard reagent is prepared by treating the aromatic halide with magnesium in ether. The Grignard reagent is sensitive to water and air and is

---

* N. L. Allinger and J. Allinger, *Structures of Organic Molecules*, in this series, pp. 50–57.

3.4 Aromatic Systems—Substitution Reactions

$$\text{C}_6\text{H}_5\text{Br} + \text{Mg} \xrightarrow{\text{Et}_2\text{O}} \text{C}_6\text{H}_5\text{-Mg-Br}$$

a Grignard reagent

usually prepared immediately before use. An important resonance form of the Grignard reagent is the dipolar form with a negative charge on the carbon atom of the organic moiety and a positive charge on the metal atom. This dipolar resonance form is useful in interpreting the reactions of Grignard reagents since the organic portion reacts with positive species

$$\left[R-Mg-Br \longleftrightarrow \overset{-}{R} \quad \overset{+}{Mg}-Br \right]$$

and potentially positive sites. Thus, a Grignard reagent reacts with the proton of water to form a C—H bond.

$$\left[R-Mg-Br \longleftrightarrow \overset{-}{R} \quad \overset{+}{Mg}-Br \right] \xrightarrow{\overset{\delta+}{H}-\overset{\delta-}{OH}} R-H + MgBr(OH)$$

More importantly, Grignard reagents react with atoms that are only slightly positive, such as the carbon atom of a carbonyl group. Treatment

$$\left[\text{>C=O} \longleftrightarrow \text{>}\overset{+}{\text{C}}-\overset{-}{\text{O}} \right]$$

of formaldehyde with phenylmagnesium bromide gives the salt of benzyl

$$\phi\text{MgBr} + \underset{H}{\overset{H}{>}}\text{C=O} \longrightarrow \phi-\underset{H}{\overset{H}{\underset{|}{\text{C}}}}-\overset{-}{\text{O}} \overset{+}{\text{MgBr}} \xrightarrow{\text{H}_2\text{O}} \phi\text{-CH}_2\text{OH} + \text{MgBr(OH)}$$

alcohol. The salt is readily hydrolyzed to benzyl alcohol. This type of reaction is an example of an addition reaction to a carbonyl group. Further examples of this type of reaction are discussed below. The following equations illustrate reactions of a Grignard reagent with other carbonyl compounds. Reactions of carbonyl compounds are discussed in greater detail in C. D. Gutsche's book in this series, *Chemistry of Carbonyl Compounds*.

$$\phi\text{-Mg-Br} + \text{O=C=O} \longrightarrow \phi-\overset{\text{O}}{\underset{\|}{\text{C}}}-\overset{-}{\text{O}} \overset{+}{\text{MgBr}} \xrightarrow{\text{H}_2\text{O}}$$

$$\phi\text{-COOH} + \text{MgBr(OH)}$$

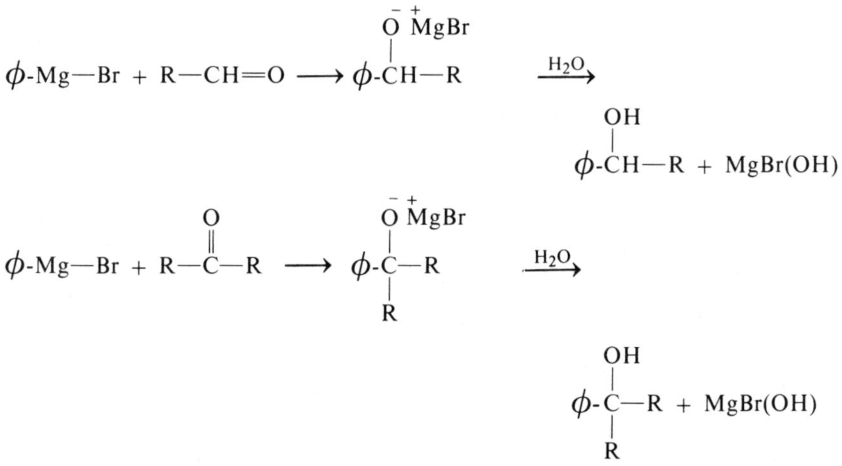

$\phi\text{-Mg}-\text{Br} + \text{R}-\text{CH}=\text{O} \longrightarrow \phi\text{-}\underset{\text{H}}{\underset{|}{\text{C}}}\text{-R} \xrightarrow{\text{H}_2\text{O}}$
$\overset{\overset{-\ +}{\text{O MgBr}}}{|}$

$\phi\text{-}\underset{\text{H}}{\underset{|}{\text{C}}}\text{-R} + \text{MgBr(OH)}$
$\overset{\text{OH}}{|}$

$\phi\text{-Mg}-\text{Br} + \text{R}-\overset{\text{O}}{\overset{\|}{\text{C}}}-\text{R} \longrightarrow \phi\text{-}\underset{\text{R}}{\underset{|}{\text{C}}}\text{-R} \xrightarrow{\text{H}_2\text{O}}$
$\overset{\overset{-\ +}{\text{O MgBr}}}{|}$

$\phi\text{-}\underset{\text{R}}{\underset{|}{\text{C}}}\text{-R} + \text{MgBr(OH)}$
$\overset{\text{OH}}{|}$

3.5 SUBSTITUTION REACTIONS OF ALIPHATIC SYSTEMS

Unlike the diazonium salt reactions, which work well only with aromatic amines, the Grignard reaction works well with aliphatic halides. For example, butyl bromide can be readily converted to valeric acid by a Grignard reaction. All of the reactions shown above for aromatic systems

$$\text{CH}_3(\text{CH}_2)_3\text{Br} \xrightarrow[\text{Et}_2\text{O}]{\text{Mg}} \text{CH}_3(\text{CH}_2)_3\text{MgBr} \xrightarrow[(2)\ \text{H}_2\text{O}]{(1)\ \text{CO}_2} \text{CH}_3(\text{CH}_2)_3\text{COOH}$$

work well for aliphatic systems. Grignard reactions are useful for preparing both aromatic and aliphatic alcohols, acids, and even deuterium labeled compounds.

$$\text{R}-\text{Mg}-\text{X} + \text{D}_2\text{O} \longrightarrow \text{R}-\text{D} + \text{MgX(OH)}$$

Nucleophilic substitution reactions constitute another type of substitution reaction that is important for certain types of aliphatic derivatives which contain a functional group that is a good leaving group. Although

$$\text{R}-\text{Y} \longrightarrow \text{R}^+ + \text{Z}^-$$

nucleophilic substitution reactions do not always involve the intermediate formation of a carbonium ion, the carbon atom to which the leaving group is attached can be thought of as being slightly positive. The *nucleophile*, which is an atom that has a pair of electrons that is seeking a

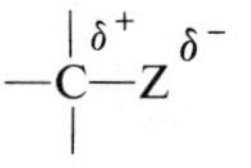

$$-\overset{|}{\underset{|}{\text{C}}}\overset{\delta+}{}-\text{Z}^{\delta-}$$

nucleus, can attack the molecule that contains Z and replace it. A specific example is the replacement of iodide from ethyl iodide by a bromide ion.

3.5 Substitution Reactions of Aliphatic Systems

Essentially any atom that has a pair of nonbonded electrons can be a

$$\text{Br}^- + \text{CH}_3-\underset{\underset{H}{|}}{\overset{\overset{H}{|}}{C}}-\text{I} \longrightarrow \text{CH}_3-\underset{\underset{H}{|}}{\overset{\overset{H}{|}}{C}}-\text{Br} + \text{I}^-$$

nucleophile. Usually these are halogen, oxygen, nitrogen, and sulfur atoms. The following are typical nucleophilic substitution reactions. Notice that arenesulfonates and alkanesulfonates, $R-SO_3^-$, are good leaving groups since they can hold a negative charge well. Reactions of

$$\text{CH}_3-\overset{..}{\text{O}}-\text{H} + \text{CH}_3\overset{\overset{\text{CH}_3}{|}}{\text{CH}}-\text{I} \longrightarrow \text{CH}_3-\text{O}-\overset{\overset{\text{CH}_3}{|}}{\text{CHCH}_3} + \text{HI}$$

$$\phi\text{-}\ddot{\text{N}}\text{H}_2 + \text{CH}_3-\text{Br} \longrightarrow \phi\text{-}\overset{\overset{H}{|}}{\text{N}}-\text{CH}_3 + \text{HBr}$$

$$\text{CH}_3\text{CH}_2\text{CH}_2\text{CH}_2\text{S}^-\text{Na}^+ + \phi\text{-CH}_2-\text{Br} \longrightarrow \text{CH}_3(\text{CH}_2)_3-\text{S}-\text{CH}_2\text{-}\phi + \text{NaBr}$$

$$\text{Cl}^- + \text{CH}_3-\text{OSO}_2-\underset{}{\bigcirc}-\text{CH}_3 \longrightarrow \text{CH}_3-\text{Cl} + \bar{\text{O}}\text{SO}_2-\underset{}{\bigcirc}-\text{CH}_3$$

$$\text{CH}_3\text{COO}^- + \underset{}{\bigcirc}-\text{O}-\text{SO}_2-\text{CH}_3 \longrightarrow \underset{}{\bigcirc}-\text{O}-\overset{\overset{O}{\|}}{\text{C}}-\text{CH}_3 + {}^-\text{OSO}_2-\text{CH}_3$$

aliphatic compounds are further discussed in W. H. Saunders' book in this series, *Ionic Aliphatic Reactions*.

Another important type of aliphatic substitution reaction involves free radicals. When a chlorine molecule absorbs light, it enters an excited state which dissociates, giving two chlorine atoms. If the chlorine is mixed with a hydrocarbon, such as ethane, a chlorine atom will abstract a hydrogen atom to form hydrogen chloride and an alkyl radical. The alkyl radical could abstract a hydrogen from another hydrocarbon molecule, but this does not result in any new products. The alkyl radical can also attack a

$$\text{CH}_3\text{CH}_3 + \text{Cl}\cdot \longrightarrow \text{CH}_3\text{CH}_2\cdot + \text{H}-\text{Cl} \qquad (3.4)$$

chlorine molecule and produce the corresponding alkyl chloride and a chlorine atom. The new chlorine atom can undergo reaction (3.4) and then reaction (3.5) can reoccur. Such a cyclic sequence of reactions is

$$\text{CH}_3\text{CH}_2\cdot + \text{Cl}-\text{Cl} \longrightarrow \text{CH}_3\text{CH}_2-\text{Cl} + \text{Cl}\cdot \qquad (3.5)$$

called a *chain reaction* and is very common in free radical reactions.

Eventually two radicals will collide and stop the reaction. However, be-

$$CH_3CH_2\cdot + Cl\cdot \longrightarrow CH_3CH_2{-}Cl$$

fore this happens, many hundred molecules of alkyl chloride can be produced, since the radical concentration is so low that there is very little chance for two radicals to meet. The overall reaction is the substitution of a hydrogen atom by a chlorine atom.

$$CH_3CH_3 \xrightarrow[Cl_2]{h\nu} CH_3CH_2Cl + HCl$$

Free radical chlorination of an alkane is a typical free radical chain reaction and like *all* free radical chain reactions can be broken down into three discrete steps: an *initiation* step which results in the formation of radicals, a *propagation* step which produces the desired product and a radical which keeps the chain reaction going, and finally a *termination* step which involves the reaction of two radicals to lead to nonradical products that are unreactive. These three steps must exist in any free radical chain reaction. The mechanism of the free radical chlorination of ethane is, then, as follows.

Initiation: $Cl_2 \xrightarrow{h\nu} 2\,Cl\cdot$

Propagation: $Cl\cdot + CH_3CH_3 \longrightarrow CH_3CH_2\cdot + HCl$

$CH_3CH_2\cdot + Cl_2 \longrightarrow CH_3CH_2Cl + Cl\cdot$

Termination: $CH_3CH_2\cdot + Cl\cdot \longrightarrow CH_3CH_2Cl$

or $CH_3CH_2\cdot + CH_3CH_2\cdot \longrightarrow CH_3CH_2CH_2CH_3$

or $Cl\cdot + Cl\cdot \longrightarrow Cl{-}Cl$

Free radical halogenation of ethane leads to ethyl chloride and so is an overall replacement of hydrogen by chlorine. What happens, however, if two or more kinds of hydrogens are present in the alkane? Which one or ones are replaced?

$$CH_3CH_2CH_3 + Cl_2 \xrightarrow{h\nu} CH_3CH_3CH_2Cl \text{ or } CH_3{-}\overset{\overset{\displaystyle Cl}{|}}{CH}{-}CH_3 \ ?$$

It turns out that chlorine atoms are very reactive. Being very reactive, they are very unselective and will consequently react with any hydrogen atom that they encounter. Thus, the free radical halogenation of propane will lead to both 1- and 2-chloropropane. However, since there are three times as many primary hydrogen atoms as secondary, to a first approximation, three times as much 1-chloropropane will be produced as 2-chloropropane.

One other point should be made about free radical halogenations. The

3.6 Substitution Reactions on Unsaturated Carbon Atoms of Aliphatic Systems

products, alkyl halides, also have hydrogen atoms and these can be attacked by chlorine atoms. However, one can minimize this reaction by using an excess of starting alkane so that the reactive chlorine atom will encounter many more alkane molecules than alkyl halide molecules. Since the molecular weight of the chlorine is much greater than that of hydrogen, it is no problem to separate the higher boiling alkyl halides from the starting alkane. Sometimes multi-chlorinated products are wanted, and in such cases an excess of the starting alkane is avoided. For example, all the halomethanes are useful and so all are produced. The mechanism for the formation of polyhaloalkanes is similar to that for the formation of haloalkanes, except that haloalkyl radicals are involved as intermediates. Free radical reactions are extensively discussed in W. A. Pryor's book in this series, *Introduction to Free Radical Chemistry*.

$$CH_4 + Cl_2 \xrightarrow{h\nu} CH_3Cl \quad \text{Methyl chloride}$$
(a useful synthetic intermediate)

CH_2Cl_2 Methylene chloride
(a useful solvent)

$CHCl_3$ Chlroform
(a useful solvent, and anesthetic)

CCl_4 Carbon tetrachloride
(a useful solvent)

3.6 SUBSTITUTION REACTIONS ON UNSATURATED CARBON ATOMS OF ALIPHATIC SYSTEMS

An extremely important functional group is the carboxylic acid group. As we will discuss below, one important characteristic of this group is

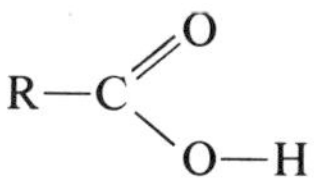

$$R-C\begin{matrix}\nearrow O \\ \searrow O-H\end{matrix}$$

that it is acidic. However, now we want to focus our attention on the many derivatives of this functional group. The most important derivatives are generated by replacing the —OH group with another atom or group of atoms.

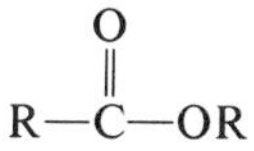

$$\underset{\|}{R-\overset{O}{C}-OR} \quad \text{Ester}$$

$$\underset{\underset{R}{|}}{R-\overset{\overset{O}{\|}}{C}-N-R} \quad \text{Amide}$$

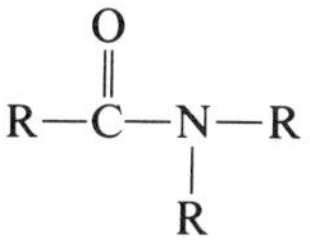

$$\text{R}-\overset{\overset{\text{O}}{\|}}{\text{C}}-\text{Cl} \qquad \text{Acid chloride}$$

$$\text{R}-\overset{\overset{\text{O}}{\|}}{\text{C}}-\text{O}-\overset{\overset{\text{O}}{\|}}{\text{C}}-\text{R} \qquad \text{Acid anhydride}$$

A typical reaction is the conversion of an acid into an ester by placing the acid into an alcoholic solution that contains a small amount of a strong acid as a catalyst. Actually, a relatively mobile equilibrium exists

$$\text{R}-\overset{\overset{\text{O}}{\|}}{\text{C}}-\text{OH} + \text{HOR} \underset{}{\overset{\text{H}^+\text{Cl}^-}{\rightleftharpoons}} \text{R}-\overset{\overset{\text{O}}{\|}}{\text{C}}-\text{OR} + \text{HOH}$$

between the acid and ester, but the ester will be present to the greatest extent if an excess of alcohol is used (which should be the case if it is used as solvent). It can be seen that this reaction is an overall substitution reaction of —OH by —OR. Indeed, mechanistically it is similar to nucleophilic reactions on saturated carbon but differs somewhat because the group to be displaced is attached to the unsaturated carbon atom of the carbonyl group.

Just as a good leaving group results in a slightly positive carbon atom, the electronegative oxygen of a carbonyl group results in a slightly positive carbon atom. This polarization is increased if the carbonyl group is

$$\left[\text{>C=O} \longleftrightarrow \text{>}\overset{+}{\text{C}}-\overset{-}{\text{O}} \right] \qquad \underset{\delta^+ \;\; \delta^-}{\text{>C=O}}$$

protonated. Consequently, the first step of the esterification of a carboxylic acid is protonation. Since the esterification is conducted in an

$$\left[\text{>C=}\underset{+}{\ddot{\text{O}}}-\text{H} \longleftrightarrow \text{>}\underset{+}{\text{C}}-\ddot{\text{O}}-\text{H} \right]$$

alcoholic solution, the proton of the strong acid is most likely attached to an alcohol molecule. The protonated carboxylic acid is now quite susceptible to attack by a nucleophile, although it can also lose its proton since proton transfer reactions are all very fast. If the protonated carboxylic

$$\text{R}-\text{C}\overset{\text{O}}{\underset{\text{OH}}{\diagdown}} + \text{H}_2\overset{+}{\text{OR}} \rightleftharpoons \left[\text{R}-\text{C}\overset{\overset{+}{\text{OH}}}{\underset{\text{OH}}{\diagdown}} \longleftrightarrow \text{R}-\text{C}\overset{\text{OH}}{\underset{+\text{OH}}{\diagdown}} \right] + \text{H}-\text{OR}$$

120   CHEMICAL INTERRELATIONS OF FUNCTIONAL GROUPS   Chap. 3

mechanisms inasmuch as a tetrahedral intermediate is formed. However, the stronger nucleophile attacks the starting carbonyl derivative directly without any activation by a proton.

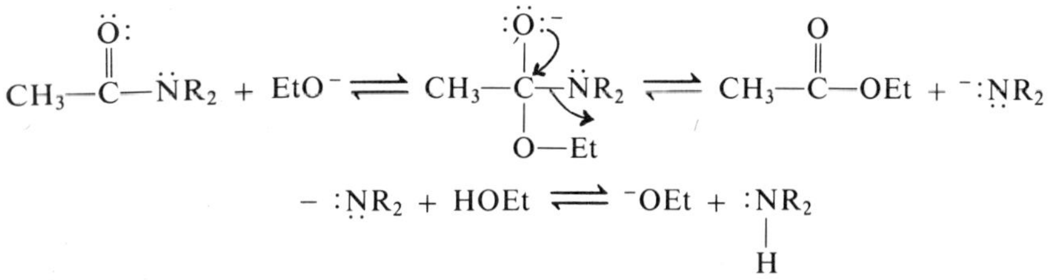

$$CH_3-\overset{\overset{\ddot{O}:}{\|}}{C}-\ddot{N}R_2 + EtO^- \rightleftharpoons CH_3-\overset{:\ddot{O}:^-}{\underset{O-Et}{C}}\ddot{N}R_2 \rightleftharpoons CH_3-\overset{O}{\overset{\|}{C}}-OEt + {}^-:NR_2$$

$$-:NR_2 + HOEt \rightleftharpoons {}^-OEt + :NR_2\!-\!H$$

Acid chlorides and anhydrides are the most reactive carboxylic acid derivatives. The following reactions represent the most common interconversions of carboxylic acid derivatives.

$$R-\overset{O}{\overset{\|}{C}}-OH + SOCl_2 \longrightarrow R-\overset{O}{\overset{\|}{C}}-Cl + SO_2 + HCl$$

$$R-\overset{O}{\overset{\|}{C}}-Cl + H_2O \longrightarrow R-\overset{O}{\overset{\|}{C}}-OH + HCl$$

$$R-\overset{O}{\overset{\|}{C}}-Cl + HOR' \longrightarrow R-\overset{O}{\overset{\|}{C}}-OR' + HCl$$

$$R-\overset{O}{\overset{\|}{C}}-Cl + H_2NR' \longrightarrow R-\overset{O}{\overset{\|}{C}}-NH-R' + HCl$$

$$R-\overset{O}{\overset{\|}{C}}-Cl + HO-\overset{O}{\overset{\|}{C}}-R' \longrightarrow R-\overset{O}{\overset{\|}{C}}-O-\overset{O}{\overset{\|}{C}}-R' + HCl$$

$$R-\overset{O}{\overset{\|}{C}}-O-\overset{O}{\overset{\|}{C}}-R + HOR' \longrightarrow R-\overset{O}{\overset{\|}{C}}-OR' + H-O-\overset{O}{\overset{\|}{C}}-R$$

$$R-\overset{O}{\overset{\|}{C}}-O-\overset{O}{\overset{\|}{C}}-R + HNR_2' \longrightarrow R-\overset{O}{\overset{\|}{C}}-NR_2' + H-O-\overset{O}{\overset{\|}{C}}-R$$

$$R-\overset{O}{\overset{\|}{C}}-OH + HOR' \underset{}{\overset{\text{acid catalyzed}}{\rightleftharpoons}} R-\overset{O}{\overset{\|}{C}}-OR' + H_2O$$

$$R-\overset{O}{\overset{\|}{C}}-OR' + {}^-OH \rightleftharpoons R-\overset{O}{\overset{\|}{C}}-O^- + HOR'$$

$$R-\overset{O}{\overset{\|}{C}}-OR' + HNR_2'' \rightleftharpoons R-\overset{O}{\overset{\|}{C}}-NR_2'' + HOR'$$

3.6 Substitution Reactions on Unsaturated Carbon Atoms of Aliphatic Systems

acid is attacked by the nucleophilic alcohol, a tetrahedral intermediate forms. Again, this reaction is reversible, since the protonated —OR

$$\left[R-C\overset{+\ \ OH}{\underset{OH}{\diagdown}} \longleftrightarrow R-C\overset{OH}{\underset{OH}{\diagdown}}{}^{+} \right] + H-\ddot{O}-R \rightleftharpoons R-\underset{\underset{H}{\overset{O}{\diagdown}}\ {}^{+}\ R}{\overset{OH}{\underset{|}{C}}}-OH$$

group is a good leaving group. Alternatively, the proton on the —OR group can be transferred to one of the —OH groups. This would most likely occur by removal of the original proton and addition of a new

$$R-\underset{\underset{H\ +\ R}{\overset{O}{\diagdown}}}{\overset{OH}{\underset{|}{C}}}-OH \underset{H_2OR^+}{\overset{HOR}{\rightleftharpoons}} R-\underset{OR}{\overset{OH}{\underset{|}{C}}}-OH \underset{HOR}{\overset{H_2OR^+}{\rightleftharpoons}} R-\underset{OR}{\overset{OH}{\underset{|}{C}}}-OH_2^+$$

proton. Now in this protonated, tetrahedral intermediate, the best leaving group is the water molecule. Again, the reaction is reversible, but the small amount of water present would require the equilibrium to lie on the

$$R-\underset{OR}{\overset{OH}{\underset{|}{C}}}\overset{+}{\frown}OH_2 \rightleftharpoons \left[R-\underset{OR}{\overset{OH}{\underset{|}{C^+}}} \longleftrightarrow R-\underset{OR}{\overset{{}^+OH}{\underset{\|}{C}}} \right] + H_2O$$

right. Loss of a proton from the protonated ester gives the neutral ester.

$$\left[R-\underset{OR}{\overset{OH}{\underset{|}{C^+}}} \longleftrightarrow R-\underset{OR}{\overset{{}^+OH}{\underset{\|}{C}}} \right] + HOR \rightleftharpoons R-\overset{O}{\underset{\|}{C}}-OR + H_2\overset{+}{O}R$$

The mechanism for the acid catalyzed hydrolysis of an ester is exactly the reverse of the above sequence, since the reaction would be carried out with an excess of water (probably as solvent) and with little alcohol present.

The main characteristic of all substitution reactions on carbonyl derivatives is the formation of a tetrahedral intermediate.

Many of these substitution reactions are base catalyzed instead of acid catalyzed. For example, the formation of an ester from an amide can be base catalyzed. These mechanisms are similar to the acid-catalyzed

$$CH_3-\overset{O}{\underset{\|}{C}}-NR_2 \xrightarrow[\text{in EtOH}]{EtO^-Na^+} CH_3-\overset{O}{\underset{\|}{C}}-OEt + HNR_2$$

3.6 Substitution Reactions on Unsaturated Carbon Atoms of Aliphatic Systems

If the hydroxyl or amine function is attached to the same molecule that contains the carboxylic acid derivative, sometimes a cyclic derivative will form. For example, γ-hydroxybutyric acid readily loses a mole of water

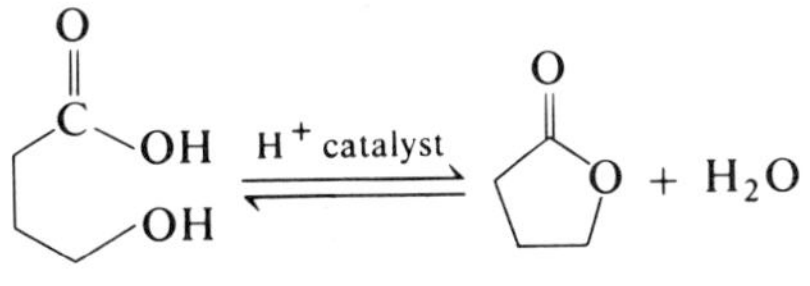

$$\text{(cyclic structure with C(=O)OH and OH)} \xrightarrow[]{H^+ \text{ catalyst}} \text{(lactone)} + H_2O$$

to form a cyclic ester, a lactone. Cyclic amides, lactams, are also readily formed. Since both functional groups are in one molecule, their inter-

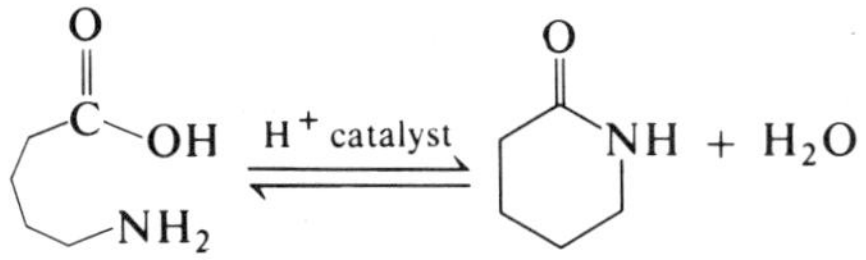

$$\text{(structure with C(=O)OH and NH}_2\text{)} \xrightarrow[]{H^+ \text{ catalyst}} \text{(lactam with NH)} + H_2O$$

action is favored by their proximity. Other aspects of the reactions, however, are the same as those of the bimolecular reactions that lead to acyclic products.

Another very common type of reaction that is related to substitution reactions of carboxylic acid derivatives is the hydrolysis of certain functional groups. For example, nitriles, or alkyl cyanides, are readily hydro-

$$R-C\equiv N \xrightarrow[H_2O]{H_3O^+} R-\underset{\underset{}{\|}}{\overset{\overset{O}{\|}}{C}}-NH_2 \xrightarrow[H_2O]{H_3O^+} R-\underset{\underset{}{\|}}{\overset{\overset{O}{\|}}{C}}-OH + NH_3$$

lyzed to amides which, of course, can be further hydrolyzed to acids. If one neglects mechanistic details, the first step of this reaction can be viewed as the addition of water across the carbon-nitrogen triple bond. A

$$H-OH + R-C\equiv N \longrightarrow R-\underset{O-H}{\overset{}{C}}=NH \longrightarrow R-\underset{\overset{\|}{O}}{\overset{}{C}}-NH_2$$

proton shift and bond migration result in an amide. Hydrolysis of the amide is the substitution of an —OH for an —NH$_2$. Since the cyanide ion, CN$^-$, is a good nucleophile, the above reaction serves as a good means of converting an alkyl halide to a carboxylic acid which contains one carbon atom more than the alkyl halide.

$$R-CH_2-Cl \xrightarrow{^-CN} R-CH_2-C\equiv N \xrightarrow[H_2O]{H_3O^+} R-CH_2-COOH$$

These reactions are further discussed in C. D. Gutsche's book in this series, *The Chemistry of Carbonyl Compounds*.

3.7 ADDITION REACTIONS TO ALDEHYDES AND KETONES

Aldehydes and ketones are two very important functional groups. Both functional groups contain a carbon-oxygen double bond, a carbonyl group. The ketone has two alkyl or aryl groups attached to the carbonyl, whereas the aldehyde has only one. Consequently, aldehydes and ketones

$$\underset{\text{an aldehyde}}{R-\overset{O}{\underset{\|}{C}}-H} \qquad \underset{\text{a ketone}}{R-\overset{O}{\underset{\|}{C}}-R}$$

have a very similar set of chemical properties. Most of the differences between reactions of aldehydes and ketones result from the greater reactivity of aldehydes due to less steric crowding. However, some differences do result from the fact that an aldehyde has a hydrogen atom instead of an alkyl group attached to the carbonyl group.

In the last section, it was pointed out that the polarity of the carbonyl group leads to attack by nucleophiles to form tetrahedral intermediates.

$$\underset{\delta^+\ \ \delta^-}{\gtrless C=O} \qquad \left[\gtrless C=O \longleftrightarrow \gtrless C^{\pm}-O^-\right]$$

These intermediates are formed by acid-catalyzed and base-catalyzed routes. For carboxylic acid derivatives, the tetrahedral intermediates react further resulting in a change of the substituent of the carbonyl group (e.g., carboxylic acids, R—CO—OH, are converted to esters, R—CO—OR). With aldehydes and ketones, there are no substituents of the carbonyl group that can be replaced, and consequently, some of the tetrahedral species are products themselves. In other words, many species can be added to aldehydes and ketones. Usually, the positive portion of the

$$\underset{R}{\overset{R}{\gtrless}}C=O + H-Z \underset{}{\overset{K_{eq}}{\rightleftarrows}} \underset{R}{\overset{R}{\gtrless}}C-O-H \\ \underset{Z}{|}$$

molecule that adds is a proton, and the carbonyl compound and the tetrahedral product exist in equilibrium. The position of the equilibrium is largely a function of the specific aldehyde or ketone, the molecule that adds, and the medium. Some common reactions are as follows.

$$\gtrless C=O + H-OH \rightleftarrows \gtrless C\underset{OH}{\overset{OH}{<}} \qquad \text{Hydrated aldehyde or ketone}$$

$$\gtrless C=O + H-OR \rightleftarrows \gtrless C\underset{OR}{\overset{OH}{<}} \qquad \text{Hemiacetal}$$

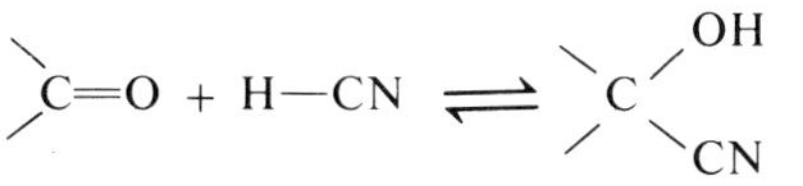

3.7 Addition Reactions to Aldehydes and Ketones

$$\ce{>C=O} + \ce{H-CN} \rightleftharpoons \ce{>C(OH)(CN)} \qquad \text{Cyanohydrin}$$

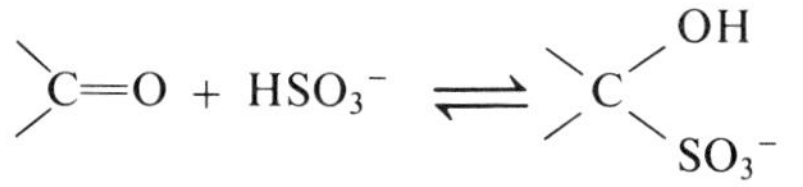

$$\ce{>C=O} + \ce{HSO_3^-} \rightleftharpoons \ce{>C(OH)(SO_3^-)} \qquad \text{Bisulfite addition product}$$

An acetal, which is formally the addition product of an ether and an aldehyde or ketone, is actually formed from the hemiacetal by the substitution of an —OH with an —OR.

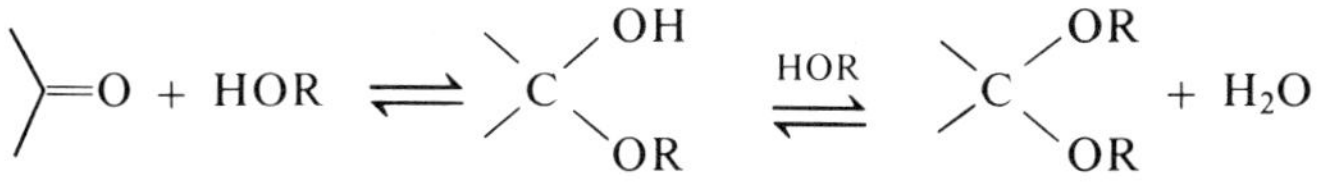

$$\ce{>C=O} + \ce{HOR} \rightleftharpoons \ce{>C(OH)(OR)} \xrightleftharpoons{\ce{HOR}} \ce{>C(OR)(OR)} + \ce{H_2O}$$

Some very important derivatives of aldehydes and ketones are formed by the initial addition of an H—NHR molecule followed by the elimination of water to form a molecule that contains a carbon-nitrogen double bond. *Oximes, phenylhydrazones, 2,4,-dinitrophenylhydrazones,* and *semi-carbazides* are four very important derivatives of aldehydes and ketones that fall into this class.

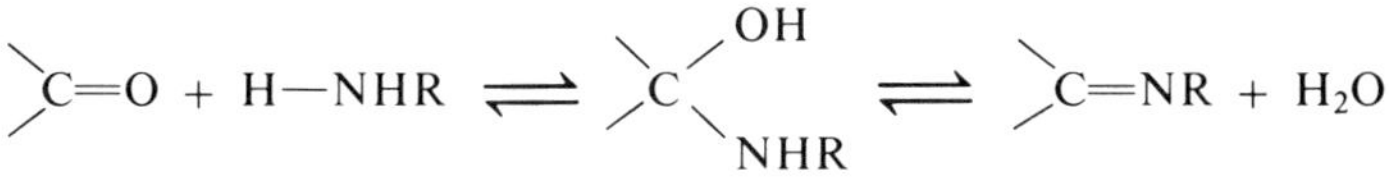

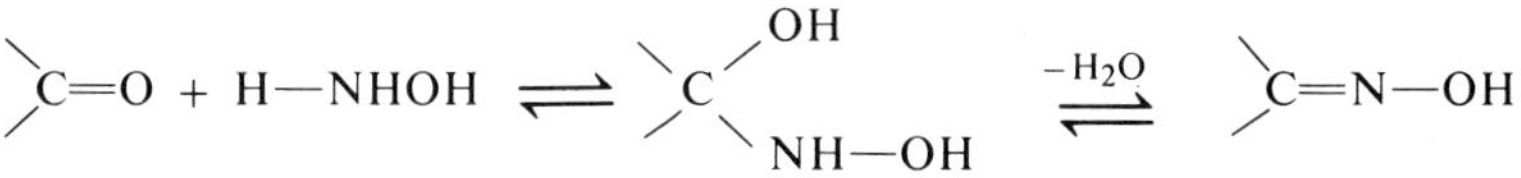

$$\ce{>C=O} + \ce{H-NHR} \rightleftharpoons \ce{>C(OH)(NHR)} \rightleftharpoons \ce{>C=NR} + \ce{H_2O}$$

$$\ce{>C=O} + \ce{H-NHOH} \rightleftharpoons \ce{>C(OH)(NH-OH)} \xrightleftharpoons{-H_2O} \ce{>C=N-OH}$$
$$\text{hydroxylamine} \qquad\qquad\qquad\qquad \text{oxime}$$

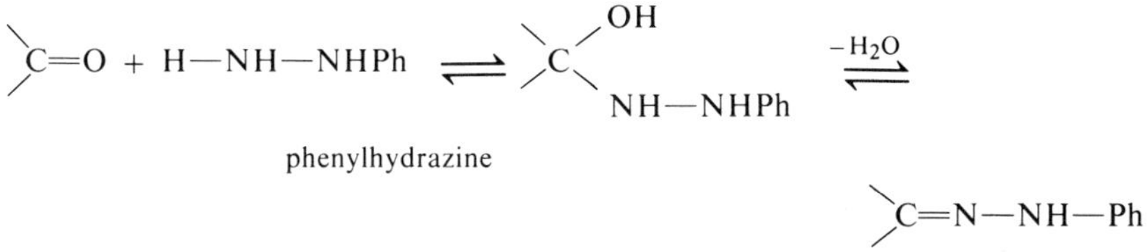

$$\ce{>C=O} + \ce{H-NH-NHPh} \rightleftharpoons \ce{>C(OH)(NH-NHPh)} \xrightleftharpoons{-H_2O}$$
$$\text{phenylhydrazine}$$

$$\ce{>C=N-NH-Ph}$$
$$\text{phenylhydrazone}$$

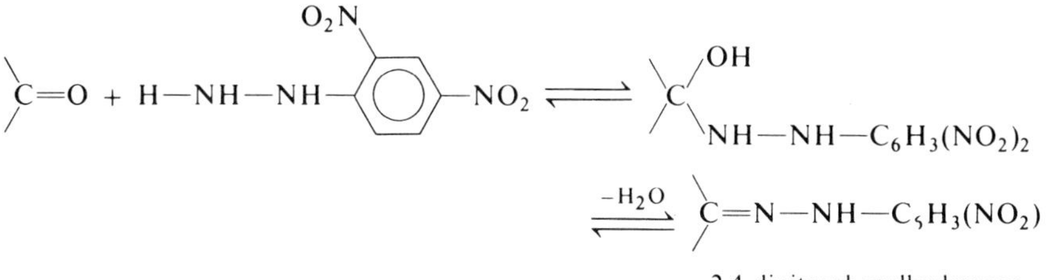

$$\ce{>C=O} + \ce{H-NH-NH-C_6H_3(NO_2)_2} \rightleftharpoons \ce{>C(OH)(NH-NH-C_6H_3(NO_2)_2)}$$

$$\xrightleftharpoons{-H_2O} \ce{>C=N-NH-C_6H_3(NO_2)_2}$$

2,4-dinitrophenylhydrazone (a 2,4-DNP)

$$\diagdown\!\!\!\!\!\diagup\!\!\text{C=O} + \text{H—NH—NH—}\overset{\overset{\displaystyle O}{\|}}{\text{C}}\text{—NH}_2 \rightleftarrows \diagdown\!\!\!\!\!\diagup\!\!\text{C}\diagdown^{\text{OH}}_{\text{NH—NH—}\overset{\overset{\displaystyle O}{\|}}{\text{C}}\text{—NH}_2}$$

semicarbazide

$$\underset{\rightleftarrows}{\overset{-\text{H}_2\text{O}}{}} \diagdown\!\!\!\!\!\diagup\!\!\text{C=N—NH—}\overset{\overset{\displaystyle O}{\|}}{\text{C}}\text{—NH}_2$$

semicarbazone

Since these derivatives of aldehydes and ketones are readily prepared, are usually solids that melt above room temperature, and are easily purified by recrystallization, they have been very useful for the identification of aldehydes and ketones. The unknown aldehyde or ketone, often a liquid, is converted to several of these derivatives. The melting points of these derivatives are determined and, by using tables of melting points of these derivatives, one can often identify the unknown aldehyde or ketone.

Another important addition reaction to carbonyl groups which leads to a functional group that contains a double bond is the *Wittig reaction.* This reaction involves the conversion of an aldehyde or ketone to an alkene by

$$\diagdown\!\!\!\!\!\diagup\!\!\text{C=O} + \bar{\text{C}}\text{H}_2\text{—}\overset{+}{\text{P}}\text{Ph}_3 \rightleftarrows \diagdown\!\!\!\!\!\diagup\!\!\text{C}\diagdown^{\text{O}^-}_{\text{CH}_2\text{—}\overset{+}{\text{P}}\text{Ph}_3} \longrightarrow \diagdown\!\!\!\!\!\diagup\!\!\text{C=CH}_2 + \bar{\text{O}}\text{—}\overset{+}{\text{P}}\text{Ph}_3$$

treating the carbonyl compound with the appropriate phosphorus ylide. The phosphorus ylide, which is a compound that contains formal positive and negative charges on adjacent atoms, is prepared from an alkyl halide and triphenylphosphine followed by reaction with an alkyl lithium reagent.

$$\text{CH}_3\text{—Br} + \text{PPh}_3 \longrightarrow \text{CH}_3\text{—}\overset{+}{\text{P}}\text{Ph}_3 + \text{Br}^- \xrightarrow{n\text{—C}_4\text{H}_9\text{Li}}$$
$$\bar{\text{C}}\text{H}_2\text{—}\overset{+}{\text{P}}\text{Ph}_3 + \text{LiBr} + n\text{—C}_4\text{H}_{10}$$

Two other very important addition reactions to aldehydes and ketones are the addition of Grignard reagents, and the addition of hydrogen. The former reaction has been discussed in Sections 3.4 and 3.5 and the latter reaction is a reduction, which will be discussed in Section 3.9.

$$\text{R—Mg—X} + \diagdown\!\!\!\!\!\diagup\!\!\text{C=O} \longrightarrow \diagdown\!\!\!\!\!\diagup\!\!\text{C}\diagdown^{\text{O}^-\overset{+}{\text{M}}\text{gX}}_{\text{R}} \xrightarrow{\text{H}_2\text{O}} \diagdown\!\!\!\!\!\diagup\!\!\text{C}\diagdown^{\text{OH}}_{\text{R}}$$

$$\text{H—H} + \diagdown\!\!\!\!\!\diagup\!\!\text{C=O} \longrightarrow \diagdown\!\!\!\!\!\diagup\!\!\text{C}\diagdown^{\text{OH}}_{\text{H}}$$

3.8 ACID-BASE REACTIONS

Reactions of aldehydes and ketones are presented in detail in C. D. Gutsche's book in this series, *The Chemistry of Carbonyl Compounds*.

3.8 ACID-BASE REACTIONS

Most acid-base reactions of organic compounds can be viewed in terms of the Brønsted acid-base theory. A Brønsted acid is defined as a proton donor and a Brønsted base is defined as a proton acceptor. Thus, the reaction of a Brønsted acid and base always results in a new Brønsted acid and base. Since proton transfer reactions are fast, an equilibrium will

$$HA + B \underset{}{\overset{K_a}{\rightleftharpoons}} HB^+ + A^-$$
$$A_1 \quad B_2 \quad\quad A_2 \quad B_1$$

be set up and will lie on the side of the weaker acid or base. It is customary to represent this equilibrium constant by K_a. The K_a (or pK_a which equals $-\log K_a$) of an acid is a measure of its acidity, since the stronger the acid, the larger will be its K_a *with respect to a given base*. Usually, the base chosen is water.

Protonation of a molecule or ion is said to convert it to its conjugate acid, and removal of a proton from a molecule or ion is said to convert it to its conjugate base. The overall electronic charges of the acids and bases are irrelevant. It should be remembered, however, that the transfer of a proton always involves the transfer of a positive charge.

One of the most characteristic aspects of a carboxylic acid is its reaction as a Brønsted acid, a proton donor. If a carboxylic acid such as acetic acid is placed in water, some of the acid molecules will transfer their protons to water and exist as acetate anions, CH_3COO^-. If a stronger

$$CH_3COOH + H_2O \overset{K_{a1}}{\rightleftharpoons} CH_3COO^- + H_3O^+$$

$$CH_3COOH + NH_3 \overset{K_{a2}}{\rightleftharpoons} CH_3COO^- + NH_4^+ \quad\quad K_{a3} > K_{a2} > K_{a1}$$

$$CH_3COOH + CH_3O^- \overset{K_{a3}}{\rightleftharpoons} CH_3COO^- + CH_3OH$$

base such as ammonia or ethoxide is added to the aqueous solution, more of the acetic acid will be converted to its conjugate base.

The reason that carboxylic acids are so acidic is that the conjugate base is a relatively stable ion. Its stability results from the fact that two good resonance structures with the negative charge on either oxygen atom

$$\left[R-C\overset{\displaystyle O}{\underset{\displaystyle O^-}{\diagdown}} \longleftrightarrow R-C\overset{\displaystyle O^-}{\underset{\displaystyle O}{\diagdown}} \right]$$

atom can be drawn. Not only is the negative charge located on an electronegative atom which can hold it well, but resonance delocalizes it.

Other types of organic structures are relatively acidic. For example, phenols and β-diketones both lose protons relatively easily. Again, the anions are resonance stabilized, with the negative charge at least partially on oxygen atoms.

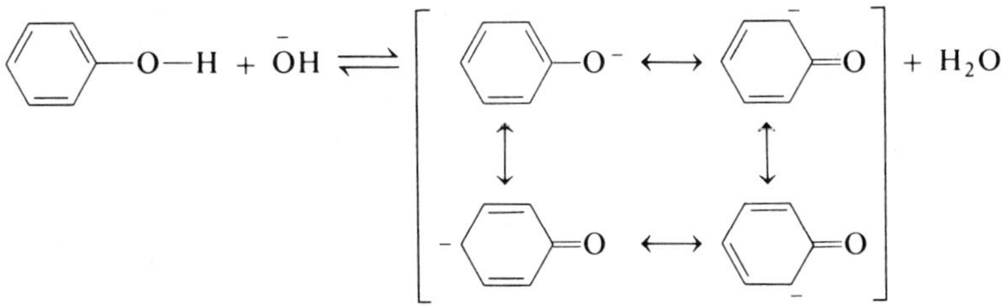

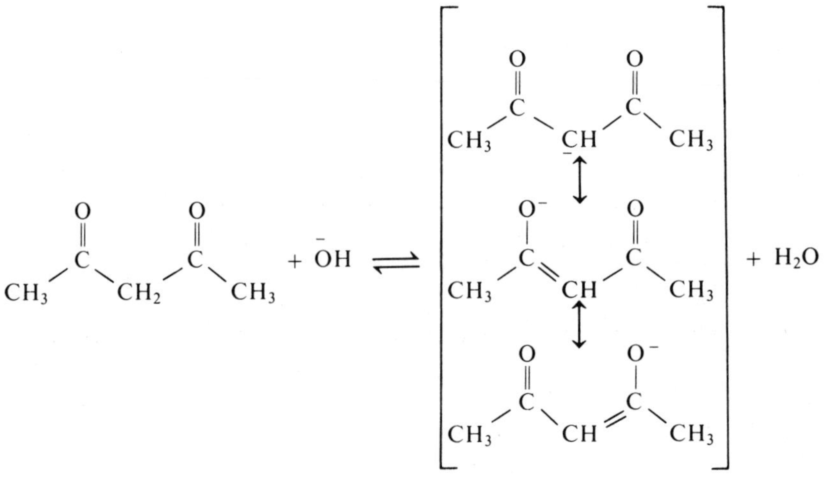

$$C_6H_5\text{—}O\text{—}H + \overline{O}H \rightleftharpoons [\text{resonance structures of phenoxide}] + H_2O$$

$$CH_3\text{—}CO\text{—}CH_2\text{—}CO\text{—}CH_3 + \overline{O}H \rightleftharpoons [\text{resonance structures of enolate}] + H_2O$$

Some organic functional groups are very basic, i.e., they tend to pick up a proton very readily. The most common basic functional group is the amino group. Just like the parent ammonia, alkylamines are basic enough to remove a proton from water, at least to a limited extent.

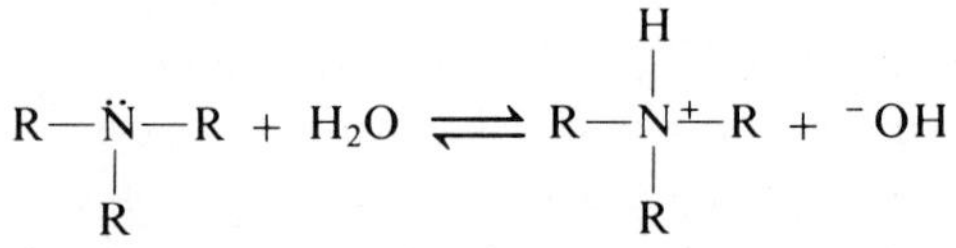

$$R\text{—}\ddot{N}R\text{—}R + H_2O \rightleftharpoons R\text{—}\overset{H}{\underset{R}{N^{\pm}}}\text{—}R + {}^-OH$$

Functional groups that contain carbonyl groups are usually weakly basic, with the basic site being the oxygen of the carbonyl group. Al-

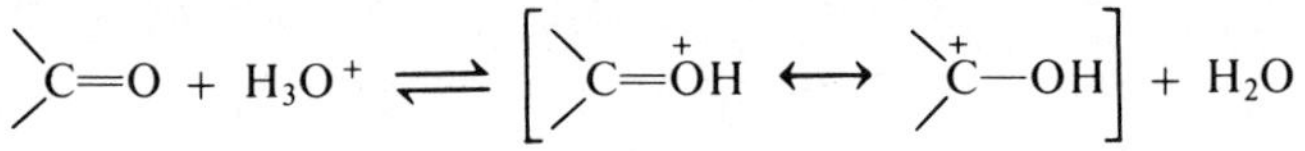

$$\text{>}C=O + H_3O^+ \rightleftharpoons \left[\text{>}C=\overset{+}{\ddot{O}}H \longleftrightarrow \text{>}\overset{+}{C}\text{—}OH\right] + H_2O$$

though only a small amount of the conjugate acid of a carbonyl-contain-

3.8 Acid-Base Reactions

ing compound may exist at equilibrium, it is often very reactive. This, of course, is the situation we encountered with the acid catalyzed esterification of carboxylic acids.

Because the K_a of an acid can be accurately measured and the structure of carboxylic acids can be varied systematically and at will, the acidity of carboxylic acids provides a beautiful means of getting at electronic and other effects of various functional groups and molecular structures. For example, the K_a's of acetic and mono-, di-, and trichloroacetic acids (Table 3-1) clearly indicate that each additional chlorine atom stabilizes

$$Cl \leftharpoondown CH_2COO^- \qquad \begin{matrix} Cl \searrow \\ Cl \nearrow \end{matrix} CH-COO^- \qquad Cl \leftharpoondown \begin{matrix} Cl \searrow \\ Cl \nearrow \end{matrix} C-COO^-$$

Table 3-1
ACIDITY OF ACETIC AND MONO-, DI-, AND TRICHLOROACETIC ACIDS

Acid	K_a
CH_3COOH	1.76×10^{-5}
$ClCH_2COOH$	155×10^{-5}
$Cl_2CHCOOH$	5140×10^{-5}
Cl_3COOH	$90{,}000 \times 10^{-5}$

the anion. This is quite consistent with the chlorine atoms inductively withdrawing electrons, as electronegative atoms should. In the discussion of the inductive effect with respect to NMR chemical shifts, it was pointed out that the inductive effect would be expected to fall off with distance. The K_a's of the acids given in Table 3-2 clearly show that the electron withdrawing effect of the chlorine becomes less important in stabilizing the carboxylate anion as it moves away from it.

The resonance effect is also readily substantiated by the K_a's of various substituted benzoic acids. The inductive effect of the nitro group should

Table 3-2
ACIDITY OF BUTYRIC AND THE CHLOROBUTYRIC ACIDS

Acid	K_a
$CH_3CH_2CH_2COOH$	1.5×10^{-5}
$CH_3CH_2\underset{\underset{Cl}{\mid}}{C}HCOOH$	139×10^{-5}
$CH_3\underset{\underset{Cl}{\mid}}{C}HCH_2COOH$	8.9×10^{-5}
$\underset{\underset{Cl}{\mid}}{C}H_2CH_2CH_2COOH$	2.96×10^{-5}

be strongly electron withdrawing and the higher acidity of *m*-nitrobenzoic acid is consistent with this. However, if only an inductive effect were operating, *p*-nitrobenzoic should be a weaker acid than the *meta* isomer,

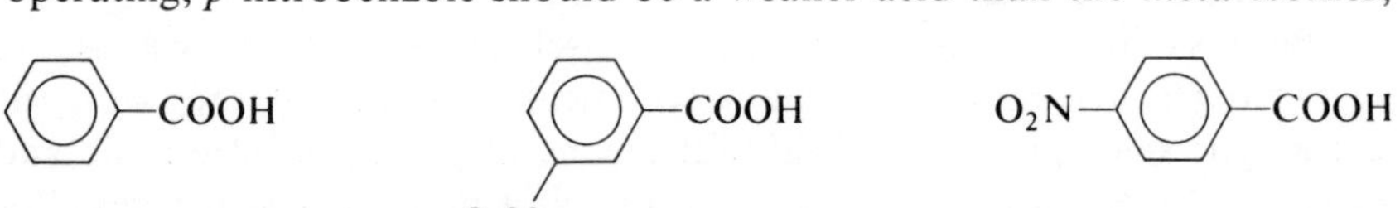

$K_a = 6.3 \times 10^{-5}$ $K_a = 34.5 \times 10^{-5}$ $K_a = 40 \times 10^{-5}$

since the nitro group is further away from the carboxylate ion in the *para* isomer. The fact that the opposite is true is consistent with a significant resonance effect of the nitro group which can delocalize electron

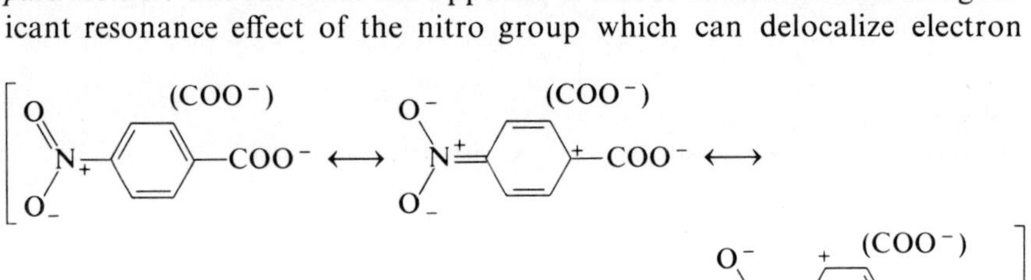

density from the carbon atom that bears the carboxylate group in the *para* isomer but not in the *meta* isomer.

The weaker acidity of *p*-methoxybenzoic acid compared to benzoic acid in spite of the electron withdrawing effect that should operate (and is

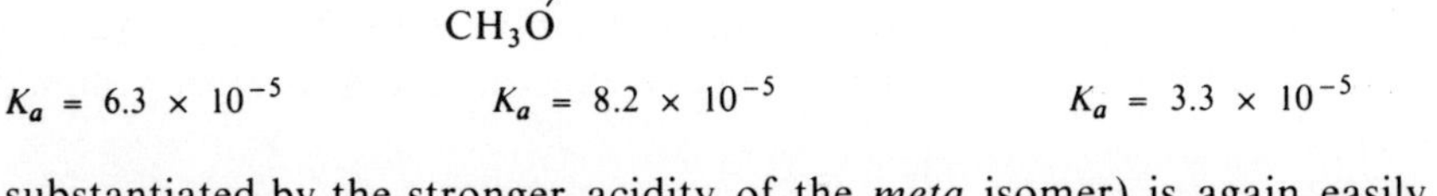

$K_a = 6.3 \times 10^{-5}$ $K_a = 8.2 \times 10^{-5}$ $K_a = 3.3 \times 10^{-5}$

substantiated by the stronger acidity of the *meta* isomer) is again easily accounted for by the resonance effect which is electron donating.

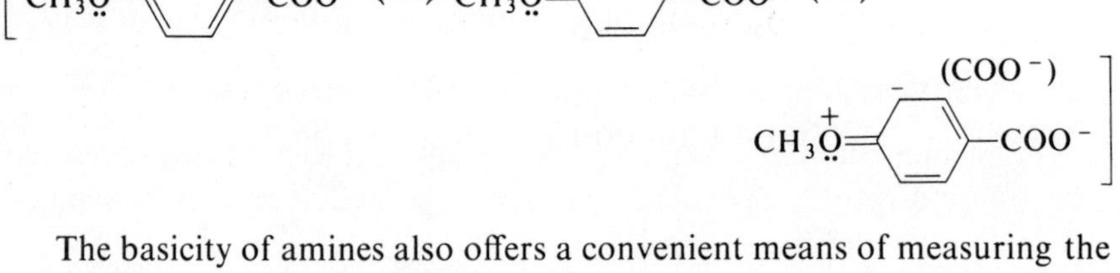

The basicity of amines also offers a convenient means of measuring the magnitude of electronic effects. In fact, since the electron pair of the

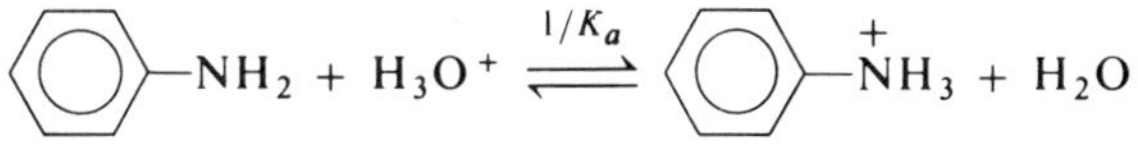

$$\text{C}_6\text{H}_5-\text{NH}_2 + \text{H}_3\text{O}^+ \underset{1/K_a}{\rightleftharpoons} \text{C}_6\text{H}_5-\overset{+}{\text{N}}\text{H}_3 + \text{H}_2\text{O}$$

nitrogen of anilines is closer to the aromatic nucleus than the oxygens of a carboxylate group, many substituent effects are larger. Moreover, resonance structures which involve this electron pair can be drawn.

In order to avoid confusion when talking about acids and bases, it is customary always to refer to acid dissociation constants. In discussions of bases, this is done by referring to the acidity of the conjugate acid of the base. When we look at a series of bases, we will therefore look at the K_a's of their conjugate acids.

Since in an aniline the nitrogen's electron pair is localized when protonated, resonance structures which delocalize it will usually decrease the aniline's basicity. The K_a's of the *meta-* and *para-*nitroanilinium ions show that this is the case.

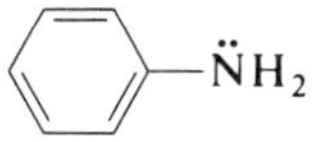

C₆H₅—N̈H₂

$K_a = 2.4 \times 10^{-3}$

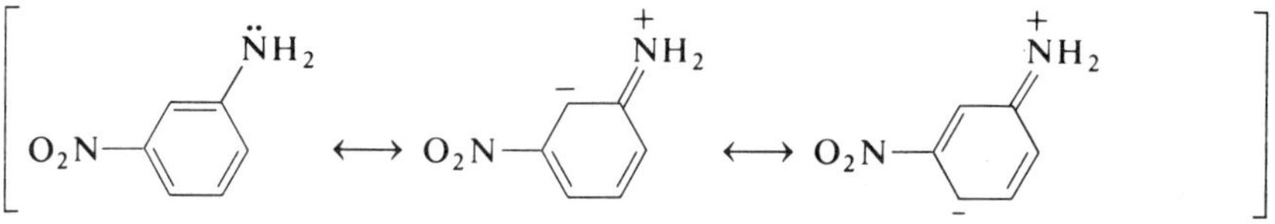

[resonance structures of m-nitroaniline: $\text{O}_2\text{N}-\text{C}_6\text{H}_4-\ddot{\text{N}}\text{H}_2 \leftrightarrow \text{O}_2\text{N}-\text{C}_6\text{H}_4-\overset{+}{\text{N}}\text{H}_2 \leftrightarrow \text{O}_2\text{N}-\text{C}_6\text{H}_4-\overset{+}{\text{N}}\text{H}_2$]

$K_a = 3.1 \times 10^{-1}$

[resonance structures of p-nitroaniline showing delocalization into ring]

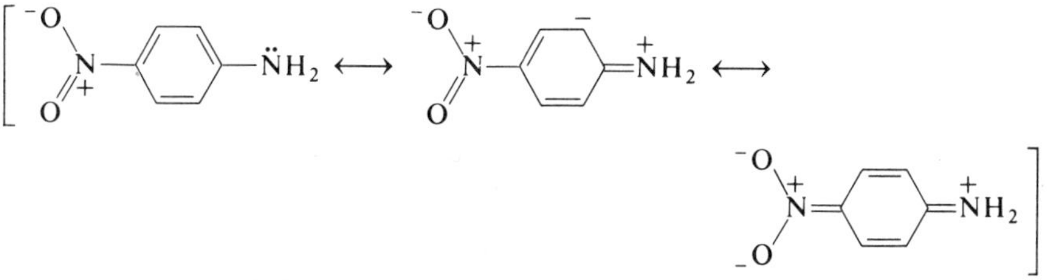

$K_a = 1.0 \times 10^{-1}$

A more extensive discussion of organic acid-base reactions can be found in R. Stewart's book in this series, *The Investigation of Organic Reactions*.

3.9 REDUCTION AND OXIDATION OF ORGANIC COMPOUNDS

In the most fundamental sense, a reduction is the gain of electrons and an oxidation is the loss of electrons. Thus, in the reaction of cerium(IV)

$$\text{Ce}^{+4} + \text{Fe}^{+2} \longrightarrow \text{Ce}^{+3} + \text{Fe}^{+3}$$

with iron(II) to give cerium(III) and iron(III), the cerium(IV) is reduced and the iron(II) is oxidized. These changes can be broken up into two half reactions and when balancing a redox reaction, it is often convenient to

$$Ce(IV) + e^- \longrightarrow Ce(III)$$
$$Fe(II) \longrightarrow Fe(III) + e^-$$

deal with half reactions. Each half reaction is balanced with respect to atoms and then each reaction is multiplied by an appropriate factor so that the sum of the two reactions will lead to no electrons on either side.

As with inorganic reactions, many important organic reactions are redox reactions. For example, the complete combustion of a hydrocarbon, a reaction very useful for the production of heat, is obviously an oxidation. Just as with inorganic reactions, the most fundamental aspect of organic redox reactions is the transfer of electrons. Organic reactions can

$$C_6H_{14} + O_2 \longrightarrow CO_2 + H_2O + \text{heat}$$

be broken up into half reactions and balanced in this fashion, although sometimes simple inspection is just as convenient. In the above reaction, simple inspection rapidly leads to the following balanced equation. How-

$$C_6H_{14} + 9\tfrac{1}{2} O_2 \longrightarrow 6\,CO_2 + 7\,H_2O$$

ever, simple inspection is not very useful for balancing the oxidation of toluene to benzoate anion with permanganate. In this case, it is easier to

$$\text{Ph-}CH_3 + MnO_4^- \xrightarrow[\text{conditions}]{\text{basic}} \text{Ph-}COO^- + MnO_2$$

balance the equation by breaking it into two half reactions. This can be done as follows:

$$\text{Ph-}CH_3 \longrightarrow \text{Ph-}COO^-$$
$$C_7H_8 \longrightarrow C_7H_5O_2^-$$
$$7\,HO^- + C_7H_8 \longrightarrow C_7H_5O_2^- + 5\,H_2O + 6\,e^- \qquad (3.6)$$
$$MnO_4^- \longrightarrow MnO_2$$
$$3\,e^- + 2\,H_2O + MnO_4^- \longrightarrow MnO_2 + 4\,HO^- \qquad (3.7)$$

The species H_2O and HO^- were used instead of H_2O and H^+, since the reaction is carried out under basic conditions. In order to get a balanced equation, Eq. (3.7) must be multiplied by 2. This yields the following upon summation of the two equations:

$$7\,HO^- + C_7H_8 + 4\,H_2O + 2\,MnO_4^-$$
$$\longrightarrow C_7H_5O_2^- + 5\,H_2O + 2\,MnO_2 + 8\,HO^-$$

3.9 Reduction and Oxidation of Organic Compounds

$$\text{C}_6\text{H}_5-\text{CH}_3 + 2\,\text{MnO}_4^- \longrightarrow \text{C}_6\text{H}_5-\text{COO}^- + 2\,\text{MnO}_2 + \text{HO}^- + \text{H}_2\text{O}$$

The oxidation of toluene to the benzoate anion may seem more like the transfer of atoms than electrons, but some organic reactions must be viewed as electron transfer reactions. For example, if naphthalene is dissolved in dimethoxyethane and treated with sodium, a dark green color appears. This green color is due to the presence of the radical anion of naphthalene formed by the reduction of napthalene by sodium. The following is a balanced equation for this reaction.

$$\text{C}_{10}\text{H}_8 + \text{Na} \longrightarrow [\text{C}_{10}\text{H}_8]^{\cdot -} + \text{Na}^+$$

In this case an electron is being added to the π-system of the organic compound and no atoms are transferred.

Although the formation of radical anions and radical cations is often encountered in modern organic chemistry and demands that one be able to think of organic redox reactions in terms of electron transfers, many reductions of organic compounds can be considered a gain of hydrogen atoms and many oxidations of organic compounds can be considered a loss of hydrogen atoms. The loss of hydrogen atoms from toluene when it is converted to benzoate anion is consistent with the fact that it is an oxidation. The addition of molecular hydrogen to an olefin or ketone

$$\text{CH}_2=\text{CH}_2 + \text{H}_2 \longrightarrow \text{CH}_3\text{CH}_3$$

$$\text{CH}_3-\overset{\overset{\displaystyle O}{\|}}{\text{C}}-\text{CH}_3 + \text{H}_2 \longrightarrow \text{CH}_3-\overset{\overset{\displaystyle OH}{|}}{\text{CH}}-\text{CH}_3$$

means that the organic molecule is reduced. These reactions can, of course, be broken up into two half reactions.

$$2\,e^- + \text{CH}_2=\text{CH}_2 + 2\,\text{H}^+ \longrightarrow \text{CH}_3-\text{CH}_3$$
$$\text{H}_2 \longrightarrow 2\,\text{H}^+ + 2\,e^-$$

$$2\,e^- + \text{CH}_3-\overset{\overset{\displaystyle O}{\|}}{\text{C}}-\text{CH}_3 + 2\,\text{H}^+ \longrightarrow \text{CH}_3-\overset{\overset{\displaystyle OH}{|}}{\text{CH}}-\text{CH}_3$$
$$\text{H}_2 \longrightarrow 2\,\text{H}^+ + 2\,e^-$$

Thus, the real or hypothetical addition or elimination of molecular hydrogen indicates a two-electron reduction or oxidation. The addition of water, however, is not a redox reaction. The following hypothetical equations illustrate how one can determine if two compounds are related by an oxidation, reduction, or nonredox reaction. In other words, the relative oxidation states of two compounds are given by these equations. It should be remembered that these equations may not represent real reactions since it may take many steps to bring about the conversion. The equations indicate, however, whether an oxidation, reduction, or nonredox reaction must be involved in the conversion.

Reductions:

$$CH_3-CH_3 + H_2 = CH_4 + CH_4$$

$$\phi\text{-COOH} + H_2 = \phi\text{-CH}=O + H_2O$$

$$HC \equiv CH + 2H_2 = CH_3-CH_3$$

Oxidations:

$$4 H_2O + CH_3-CH_3 = CO_2 + CO_2 + 7 H_2$$

$$2 H_2O + CH_2=CH_2 = \underset{\underset{OH}{|}}{CH_2}-\underset{\underset{OH}{|}}{CH_2} + H_2$$

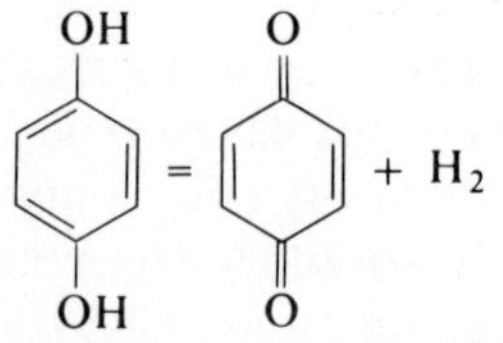

$$\text{hydroquinone} = \text{quinone} + H_2$$

$$H_2O + CH_3CH_2OH = CH_3COOH + 2 H_2$$

Neither Reduction or Oxidation:

$$CH_2=CH_2 + H_2O = CH_3-CH_2OH$$

$$CH_3-CH_3 + H_2O = CH_4 + CH_3OH$$

$$R-C \equiv N + H_2O = R-\overset{\overset{O}{\|}}{C}-NH_2$$

$$CH_3-O-CH_3 + H_2O = CH_3OH + HOCH_3$$

From the above equations, the student should realize that certain functional groups are interrelated by definite oxidation changes. Since most stable organic molecules are diamagnetic, most functional groups are related by two-electron changes. The student should be aware of the various oxidation states of various functional groups since he will then know what type of reaction (with respect to reduction and oxidation) will be needed to interconvert them. The following series are very common. Each conversion is accompanied by a two-electron change.

3.9 Reduction and Oxidation of Organic Compounds

$$R-CH_3 + H_2O \longrightarrow R-CH_2OH + 2H^+ + 2e^-$$
$$R-CH_2OH \longrightarrow R-CH=O + 2H^+ + 2e^-$$
$$R-CH=O + H_2O \longrightarrow R-COOH + 2H^+ + 2e^-$$
$$R-COOH + H_2O \longrightarrow R-OH + CO_2 + 2H^+ + 2e^-$$

$$CH_4 + CH_4 \longrightarrow CH_3-CH_3 + 2H^+ + 2e^-$$
$$CH_3-CH_3 \longrightarrow CH_2=CH_2 + 2H^+ + 2e^-$$
$$CH_2=CH_2 \longrightarrow CH\equiv CH + 2H^+ + 2e^-$$

$$R-NH_2 + H_2O \longrightarrow R-NH-OH + 2H^+ + 2e^-$$
$$R-NH-OH \longrightarrow R-N=O + 2H^+ + 2e^-$$
$$R-N=O + H_2O \longrightarrow R-NO_2 + 2H^+ + 2e^-$$

The above series includes only carbon, oxygen, and nitrogen containing compounds and, indeed, these are the most common types of compounds known. Substituents which can be hypothetically replaced with —OH groups by hydrolysis are in the same oxidation state as the compound that would have an —OH in place of the substituent. For example, methyl bromide and methanol are in the same oxidation state since the two can be interconverted without any reductions or oxidations taking

$$CH_3-Br + H_2O \longrightarrow CH_3-OH + H-Br$$

place. In a similar fashion, 1,1,1-trichloroethane and acetic acid are in the same oxidation state.

$$CH_3CCl_3 + 2H_2O \longrightarrow CH_3COOH + 3HCl$$

Further consideration of this point should make it clear that the addition of bromine to an olefin is really an oxidation of the olefin, since the

$$CH_2=CH_2 + Br_2 \longrightarrow \underset{\underset{Br}{|}}{CH_2}-\underset{\underset{Br}{|}}{CH_2}$$

$$\underset{\underset{Br}{|}}{CH_2}-\underset{\underset{Br}{|}}{CH_2} + 2H_2O \longrightarrow \underset{\underset{OH}{|}}{CH_2}-\underset{\underset{OH}{|}}{CH_2} + 2HBr$$

$$CH_2=CH_2 + 2H_2O \longrightarrow \underset{\underset{OH}{|}}{CH_2}-\underset{\underset{OH}{|}}{CH_2} + 2H^+ + 2e^-$$

conversion of an olefin to a glycol is a two-electron oxidation. Also, it should be apparent that the bromine is being reduced to bromide.

Even though the student is aware of the oxidation state of an organic compound, he must also know specific means of reducing or oxidizing the compound, since different reagents and conditions will often lead to different products. Many reductions can be brought about by catalytic hydrogenations. A very useful reductant in the laboratory is lithium aluminum

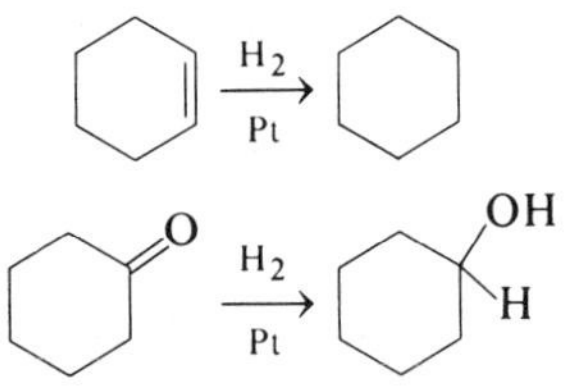

$$\bigcirc \xrightarrow{H_2}{Pt} \bigcirc$$

$$\bigcirc\!\!=\!\!O \xrightarrow{H_2}{Pt} \bigcirc\!\!\!\begin{smallmatrix}OH\\H\end{smallmatrix}$$

hydride, LiAlH$_4$. In general, lithium aluminum hydride will reduce a polarized multiple bond, but will leave nonpolarized multiple bonds and single bonds untouched. The following reactions illustrate its usefulness. The reductions are usually followed by an acidic hydrolysis.

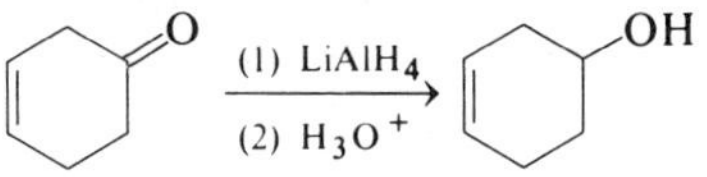

$$\bigcirc\!\!=\!\!O \xrightarrow[(2)\ H_3O^+]{(1)\ LiAlH_4} \bigcirc\!\!-\!\!OH$$

$$R-COOCH_3 \xrightarrow[(2)\ H_3O^+]{(1)\ LiAlH_4} R-CH_2OH + HOCH_3$$

$$R-C\!\!\equiv\!\!N \xrightarrow[(2)\ H_3O^+]{(1)\ LiAlH_4} R-CH_2-NH_2$$

$$R-\overset{O}{\overset{\|}{C}}-NH_2 \xrightarrow[(2)\ H_3O^+]{(1)\ LiAlH_4} R-CH_2-NH_2$$

Often special reductants are used to bring about certain conversions. For example, nitro compounds are readily reduced to amines with tin in the presence of acid. Zinc amalgam and hydrochloric acid is used

$$R-NO_2 \xrightarrow[HCl]{Sn} R-NH_2$$

to reduce the carbonyl group of aldehydes and ketones to a methylene. This reaction is called the *Clemmensen reaction*. The same conversion can

$$R-\overset{O}{\overset{\|}{C}}-R \xrightarrow[HCl]{Zn-Hg} R-CH_2-R \text{ (Clemmensen)}$$

be brought about by the *Wolff-Kishner reaction*, which involves heating the ketone in the presence of hydrazine and base.

$$R-\overset{O}{\overset{\|}{C}}-R \xrightarrow[^-OH]{N_2H_4} R-CH_2-R \text{ (Wolff-Kishner)}$$

3.9 Reduction and Oxidation of Organic Compounds

Many laboratory oxidations of organic compounds can be carried out by chromic acid or permanganate ion. Other special reagents are useful

$$CH_3CH_2CH_2CH_2OH \xrightarrow{MnO_4^-} CH_3CH_2CH_2COOH$$

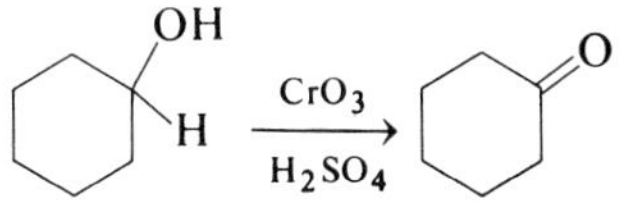

$$\text{cyclohexanol} \xrightarrow[H_2SO_4]{CrO_3} \text{cyclohexanone}$$

for special conversions. For example, it is difficult to stop at the aldehyde stage of oxidation since aldehydes are easily reduced to alcohols or oxidized to acids. However, cerium(IV) readily converts benzyl alcohols to

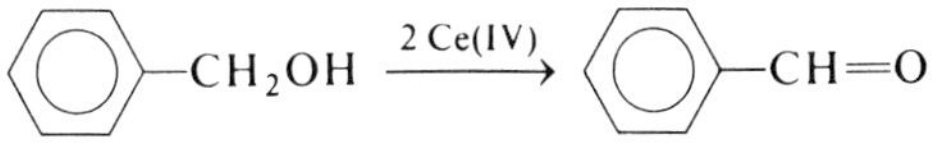

$$Ph\text{-}CH_2OH \xrightarrow{2\,Ce(IV)} Ph\text{-}CH=O$$

benzaldehydes. Other specific oxidations are the following.

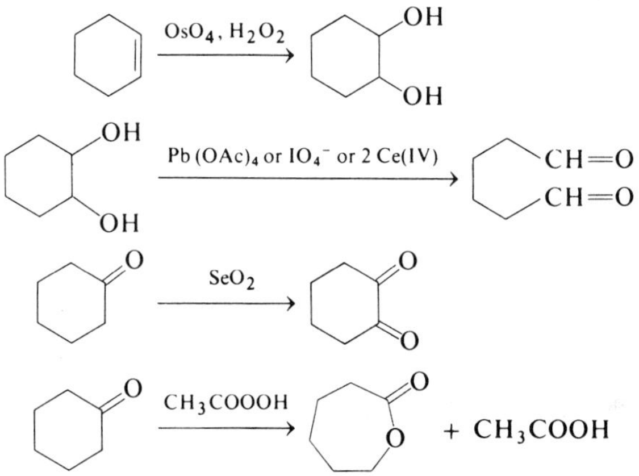

$$\text{cyclohexene} \xrightarrow{OsO_4, H_2O_2} \text{trans-1,2-cyclohexanediol}$$

$$\text{1,2-cyclohexanediol} \xrightarrow{Pb(OAc)_4 \text{ or } IO_4^- \text{ or } 2\,Ce(IV)} \text{adipaldehyde (CH=O, CH=O)}$$

$$\text{cyclohexanone} \xrightarrow{SeO_2} \text{1,2-cyclohexanedione}$$

$$\text{cyclohexanone} \xrightarrow{CH_3COOOH} \text{caprolactone} + CH_3COOH$$

Another very important type of reaction that involves oxidation is degradation. A degradation reaction involves the removal of one or more carbon atoms from the molecule. The *Hofmann* degradation of amides is a good example of such a reaction. The following is a balanced equation for this reaction.

$$H_2O + CH_3CH_2CH_2\overset{O}{\overset{\|}{C}}-NH_2 + Br_2 \xrightarrow{^-OH} CH_3CH_2CH_2-NH_2 + 2\,HBr + CO_2$$

Another very important degradation reaction is the conversion of methyl ketones to iodoform and the carboxylic acid that contains one less carbon atom by treating the ketone with iodine and base. This reaction is called the *iodoform reaction* and is used as a qualitative test for methyl

136 CHEMICAL INTERRELATIONS OF FUNCTIONAL GROUPS Chap. 3

$$R-\overset{O}{\underset{\|}{C}}-CH_3 \xrightarrow[-OH]{I_2} R-COO^- + HCI_3 \xrightarrow{H^+} R-COOH$$
(yellow solid)

ketones since the formation of iodoform, which is a yellow precipitate, can be detected readily. The reaction is also a synthetically useful way of preparing carboxylic acids.

Ozonolysis reactions are degradations that often involve the loss of several carbon atoms. These reactions were discussed under addition reactions. Organic redox reactions are further discussed in K. L. Rinehart's book in this series, *Oxidation and Reduction of Organic Compounds*.

PROBLEMS

1. Give the major organic product or products expected from the following reactions.

(a) 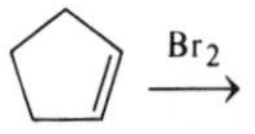 cyclopentene $\xrightarrow{Br_2}$

(b) 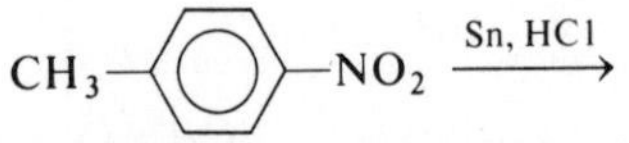 $CH_3-\langle \bigcirc \rangle-NO_2 \xrightarrow{Sn, HCl}$

(c) 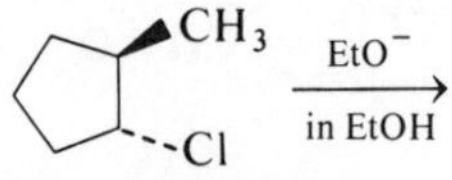 cyclopentane with CH₃ and Cl substituents $\xrightarrow[\text{in EtOH}]{EtO^-}$

(d) 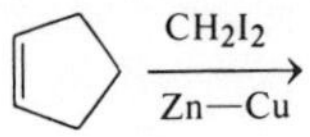 cyclopentene $\xrightarrow{CH_2I_2}{Zn-Cu}$

(e) 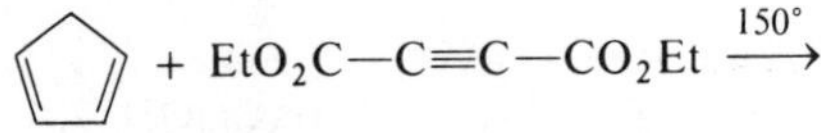 cyclopentadiene $+ EtO_2C-C\equiv C-CO_2Et \xrightarrow{150°}$

(f) 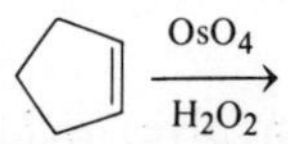 cyclopentene $\xrightarrow[H_2O_2]{OsO_4}$

(g) Ph–CH₂CH=O $\xrightarrow[Pt, 25°]{H_2}$

(h) cyclohexanone $\xrightarrow{CF_3COOOH}$

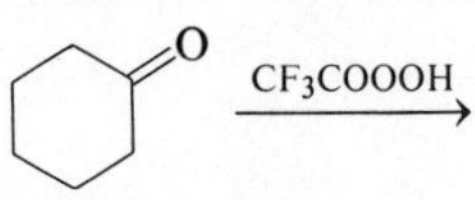

(i) 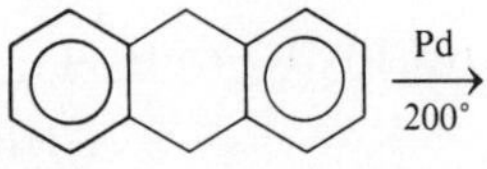 9,10-dihydroanthracene $\xrightarrow[200°]{Pd}$

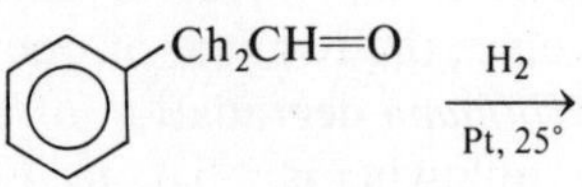

(j) $Ph-CH_2-CH=O + H_2N-NH-\overset{O}{\underset{\|}{C}}-NH_2 \xrightarrow[\text{of } H^+]{trace}$

Problems

(k) 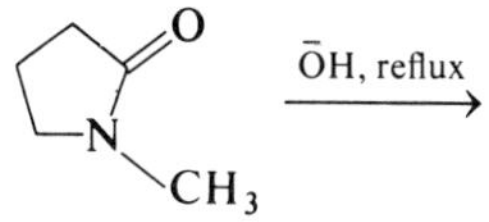 [1-methyl-2-pyrrolidinone] $\xrightarrow{\bar{O}H, \text{reflux}}$

(l) Ph—NH—CH$_3$ $\xrightarrow{(CH_3CO)_2O}$

(m)

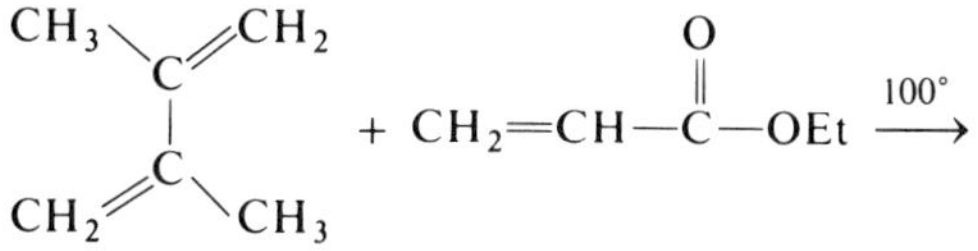

$$\underset{CH_2}{\overset{CH_3}{\underset{\|}{C}}}\!\!-\!\!\underset{CH_3}{\overset{CH_2}{\underset{\|}{C}}} + CH_2=CH-\overset{O}{\underset{\|}{C}}-OEt \xrightarrow{100°}$$

(n) 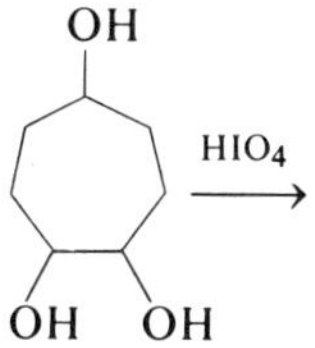 [cycloheptane-1,3,4-triol] $\xrightarrow{HIO_4}$

(o) [2-methylcyclopentanone] + Ph$_3\overset{+}{P}-\overset{-}{C}H_2 \longrightarrow$

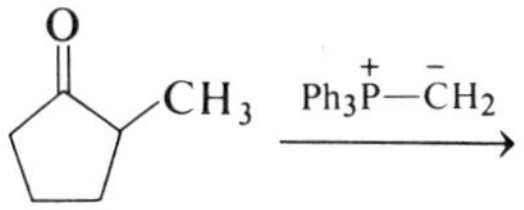

(p) $(CH_3CH_2)_3\overset{+}{N}-\overset{-}{O} \xrightarrow{150°}$

(q) 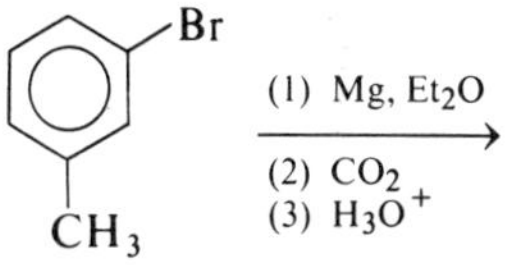 [3-bromotoluene] $\xrightarrow[\text{(2) CO}_2]{\text{(1) Mg, Et}_2\text{O}}\xrightarrow{\text{(3) H}_3\text{O}^+}$

(r) Ph—CO—CH$_2$CH$_3$ $\xrightarrow[\text{HCl}]{\text{Zn—Hg}}$

(s) 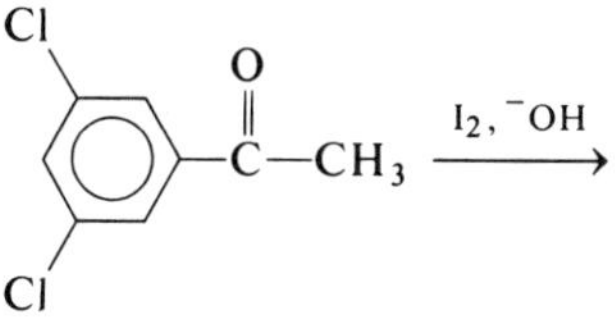 [3,5-dichloroacetophenone] $\xrightarrow{I_2, \,^-OH}$

(t) Ph—Br $\xrightarrow[\text{(2) CH}_3\text{—CO—CH}_3]{\text{(1) Mg, Et}_2\text{O}}$ (3) H$_3$O$^+$

(u) 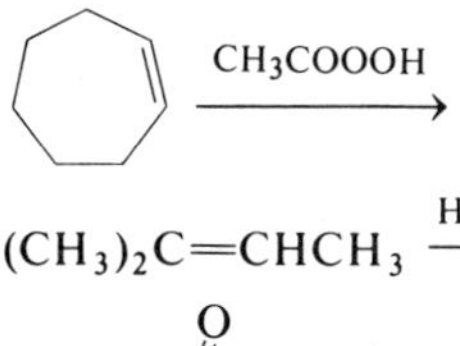 [cycloheptene] $\xrightarrow{CH_3COOOH}$

(v) (CH$_3$)$_2$C=CHCH$_3$ $\xrightarrow{HCl}$

(w) 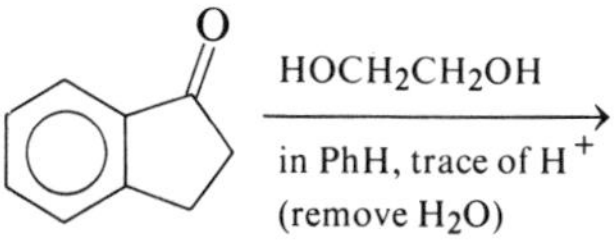 [indan-1-one] $\xrightarrow[\text{in PhH, trace of H}^+]{\text{HOCH}_2\text{CH}_2\text{OH}}$ (remove H$_2$O)

(x) $CH_3(CH_2)_4CH=CH_2 \xrightarrow{\text{small amount of } H_2SO_4 \text{ in } H_2O}$

(y) $CH_3(CH_2)_4CH=CH_2 \xrightarrow{\text{(1) } BH_3}_{\text{(2) } H_2O_2}$

(z) cyclopentyl–Br $\xrightarrow{\text{(1) Mg, Et}_2\text{O}}_{\text{(2) } CH_3-CH=O \atop \text{(3) } H_3O^+}$

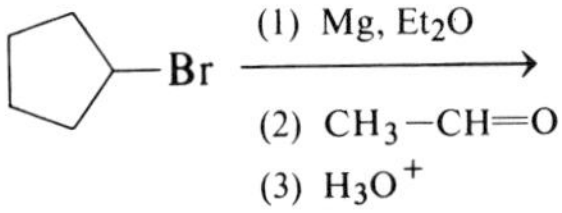

(aa) $(CH_3)_2CH-CH=O \xrightarrow{\text{(1) Ph-Mg-Br}}_{\text{(2) } H_3O^+}$

(ab) $CH_2=CHCH_2CH_2CH_2Br \xrightarrow{\text{(1) } K^+\bar{C}N}$

(ac) $(CH_3)_2CH(CH_2)_2COOH \xrightarrow{\text{(1) LiAlH}_4}_{\text{(2) } H_3O^+}$

(ad) $NC-\text{C}_6\text{H}_4-NH_2 \xrightarrow{\text{(1) HONO, 0°}}_{\text{(2) CuBr}}$

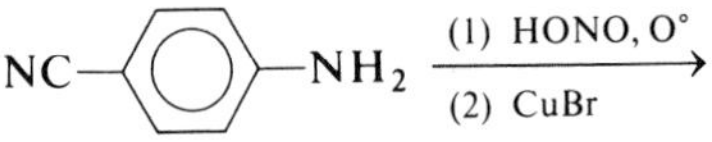

(ae) cyclopentyl-C(=O)-NH$_2$ $\xrightarrow{Br_2, \,^-OH}$

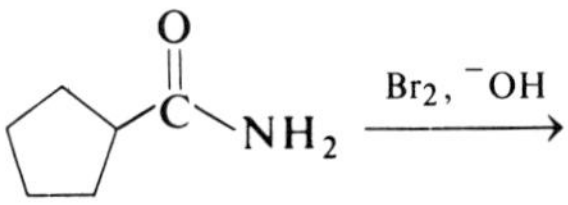

(af) cyclohexyl-CH$_2$COOEt $\xrightarrow{\text{(1) } H_3O^+, H_2O \atop \text{(2) SOCl}_2 \atop \text{(3) HNEt}_2}$

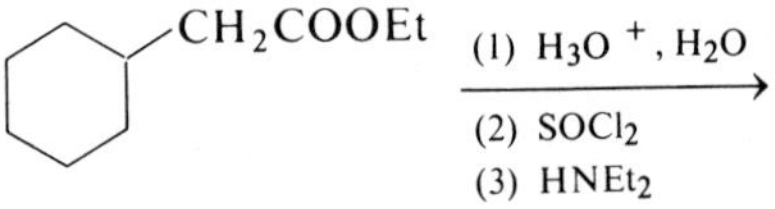

(ag) $CH_3CH_2CH_2OH \xrightarrow{HBr}$

(ah) Ph-C(=O)-Ph $\xrightarrow{N_2H_4, \,^-OH}_{190°}$

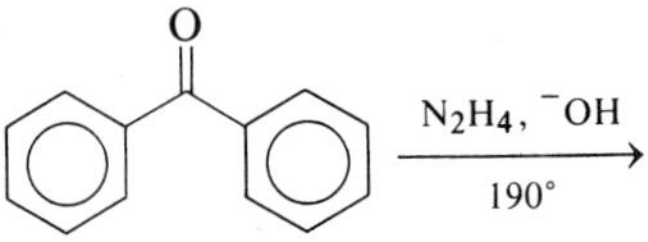

(ai) $CH_3(CH_2)_3CH=O \xrightarrow{HCN}$

(aj) cyclopentanone $\xrightarrow{H_2N-NH-Ph}$

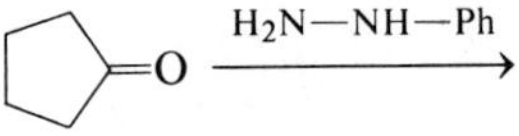

(ak) benzene $\xrightarrow{Br_2}_{FeBr_3}$

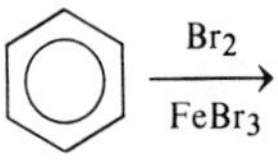

(al) $CH_3-\text{C}_6\text{H}_4-CH_3 \xrightarrow{(CH_3)_2CH-Cl}_{AlCl_3}$

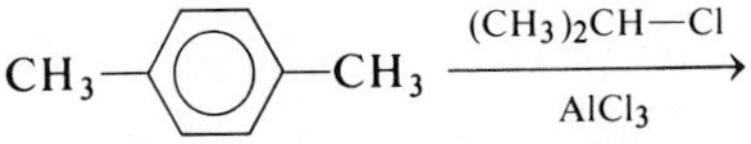

(am) $CH_3CH_2C{\equiv}CCH_3 \xrightarrow{H_2 \text{ (1 equiv.)}}_{\text{Pd (poisoned)}}$

(an)
$$CH_3CHCHCH_2CH_3 \xrightarrow[(2)\ D_2O]{(1)\ Mg,\ Et_2O}$$
with CH$_3$ on the second carbon and Br on the third carbon

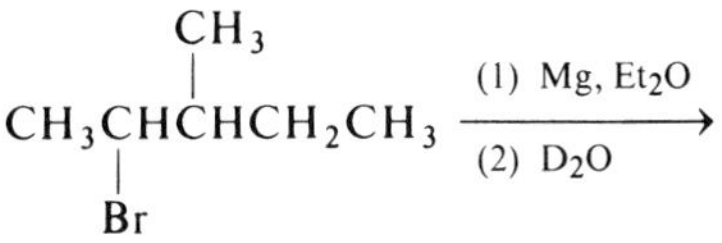

(ao)
$$CH_3CH_2CHCH_2CH_3 \xrightarrow{KOH}$$
with Br on the middle carbon

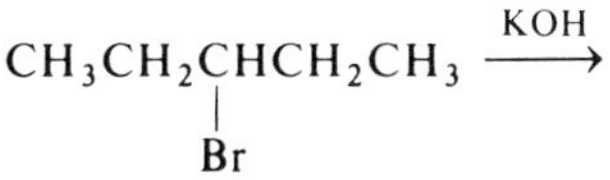

(ap)
benzene $\xrightarrow[AlCl_3]{CH_3CH_2\overset{O}{\overset{\|}{C}}-Cl}$

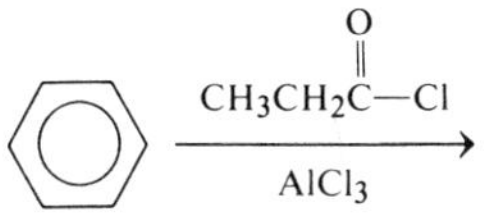

(aq)
$$CH_3CH_2CH-CH-CH_2 \xrightarrow{Zn}$$
with CH$_3$, Br, Br substituents

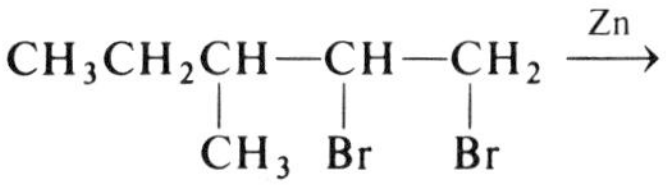

(ar) cyclopentanone + 2,4-dinitrophenylhydrazine $\xrightarrow{\text{trace of } H^+}$

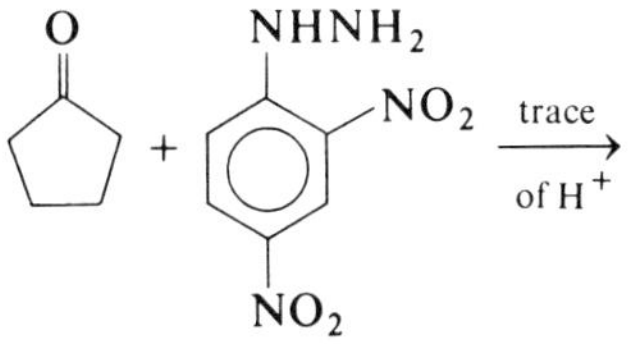

(as) cyclopentyl—COOH $\xrightarrow[(2)\ CH_3CH_2OH]{(1)\ SOCl_2}$

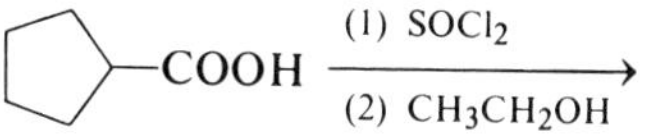

(at) $(CH_3)_3CCH_2CH_2COOCH_3 \xrightarrow[CH_3CH_2CH_2OH]{\text{trace of } H^+}$

(au) 1,2-dihydronaphthalene $\xrightarrow[(2)\ H_2O_2]{(1)\ BH_3}$

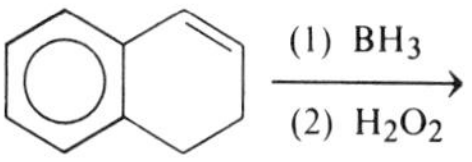

(av) $CH_3(CH_2)_5OH \xrightarrow{CrO_3,\ H_2SO_4}$

(aw) 4-chlorobenzyl alcohol $\xrightarrow{2\ Ce(IV)}$

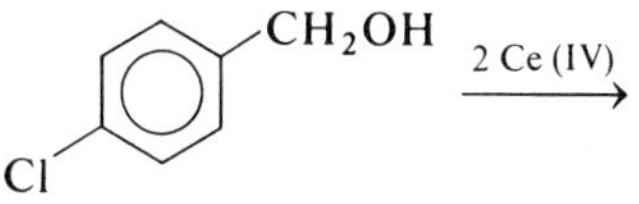

(ax) PhCH$_2$Br $\xrightarrow[(3)\ H_3O^+]{(1)\ Mg,\ Et_2O \atop (2)\ CH_3CH=O}$

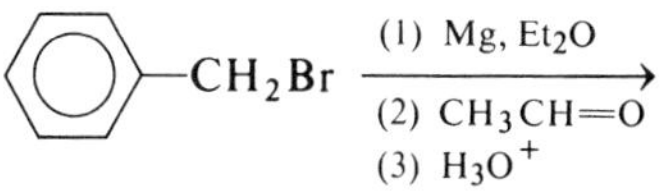

(ay) δ-valerolactam $\xrightarrow[(2)\ H_3O^+]{(1)\ LiAlH_4}$

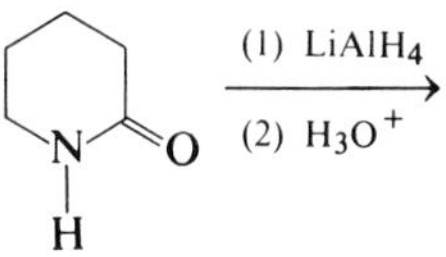

(az) cyclopentyl-CH$_2$CH$_2$—OSO$_2$Ph $\xrightarrow{K^+\ {}^-CN}$

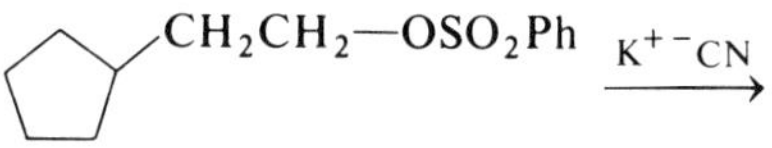

2. Draw the four isomeric monochlorides that could be obtained from the free radical chlorinations of 3-methylpentane. If 1°, 2°, and 3° hydrogen atoms are attacked at equal rates, what would be the approximate relative yields of these four alkyl chlorides?

3. Balance the following equations.

(a) C_6H_{10} + MnO$_4$ $\xrightarrow[\text{H}_2\text{O}]{-\text{OH}}$ cyclohexane-1,2-dicarboxylate (COO$^-$, COO$^-$) + MnO$_2$

(b) Ph—CH$_2$OH + CrO$_4^=$ $\xrightarrow[\text{H}_2\text{O}]{\text{H}_3\text{O}^+}$ Ph—COOH + Cr(III)

4. Draw the structures of all the possible products that could be obtained by the mononitration of the following aromatic compounds.

(a) *ortho*-Dibromobenzene

(b) *meta*-Chlorobenzoic acid

(c) *para*-Chlorotoluene

(d) 1,2,3-Trimethylbenzene

(e) 1,3,5-Trimethylbenzene

(f) *ortho*-Bromochlorobenzene

(g) *meta*-Bromochlorobenzene

(h) *para*-Bromochlorobenzene

5. Compound **A**, $C_8H_{17}Br$, was treated with NaOEt in EtOH to give olefin **B**, C_8H_{16}. Olefin **B** was treated with ozone followed by an oxidative workup to give only one product, **C**. Compound **C** took up one mole of hydrogen when catalytically hydrogenated to give **D**, $C_4H_{10}O$, which was an alcohol. Treatment of alcohol **D** with sulfuric acid gave three olefins, **E**, **F**, and **G**, with formulas C_4H_8. Give the structures for **A** to **G**.

6. Compound **H**, C_7H_{12}, when catalytically hydrogenated absorbed two moles of hydrogen. Ozonolysis of 0.1 mole of **H** followed by a reductive workup gave 0.2 mole of CH$_2$=O and 0.1 mole of O=CHCH$_2$CH$_2$COCH$_3$. Draw the structure of **H**.

7. Hydrocarbon **I**, $C_{10}H_{12}$, took up four moles of hydrogen upon catalytic hydrogenation. When **I** was oxidized with KMnO$_4$ a high yield of a benzenedicarboxylic acid, **J**, was obtained. The diacid **J** was reduced to a xylene which gave three mononitro compounds when treated with a nitric acid-sulfuric acid mixture. Ozonolysis of **I** followed by an oxidative workup gave **K** and acetic acid. Give the structures for **I** to **K**.

8. Give reasonable laboratory synthetic schemes for the following conversions. Use any inorganic or organic reagents, any organic compound having three or less carbon atoms, or benzene as reactants in addition to the compound given.

(a) 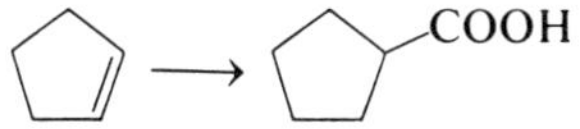 cyclopentene ⟶ cyclopentane-COOH

(b) 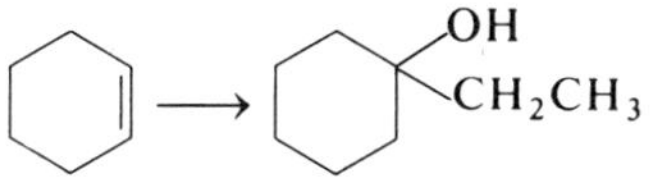 cyclohexene ⟶ 1-ethylcyclohexanol (OH, CH$_2$CH$_3$)

(c) $CH_3CH_2CH_2Br \longrightarrow CH_3CH_2CH_2CH_2NH_2$

(d) 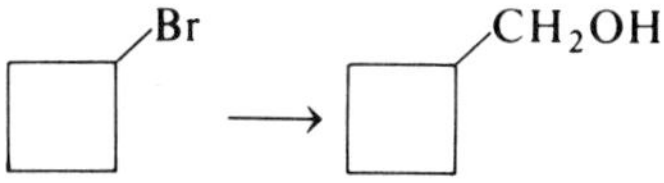 bromocyclobutane ⟶ (hydroxymethyl)cyclobutane

(e) 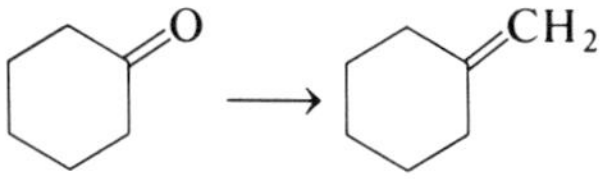 cyclohexanone ⟶ methylenecyclohexane

(f) 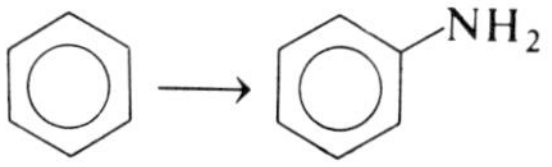 benzene ⟶ aniline (NH_2)

(g) 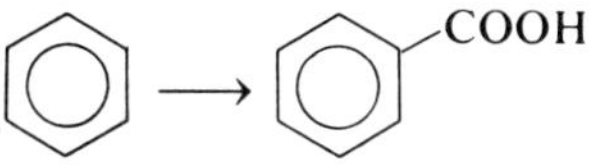 benzene ⟶ benzoic acid (COOH)

(h)  cyclopentene ⟶ $NH_2(CH_2)_7NH_2$

(i) $(CH_3)_3CCH_2COOH \longrightarrow (CH_3)_3CCH_2CH_2N(CH_2CH_3)_2$

(j) $CH_3CH_2CH_2COOCH_2CH_3 \longrightarrow CH_3CH_2CH_2CH_2COOH$

(k) 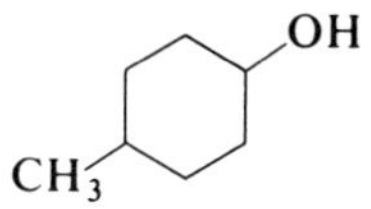 4-methylcyclohexanol ⟶ $O{=}CHCH_2CH_2\overset{\underset{\displaystyle CH_3}{|}}{C}HCH_2CH{=}O$

(l) $CH_3CH_2CH{=}CH_2 \longrightarrow Ph{-}\overset{\underset{\displaystyle CH_3}{|}}{C}HCH_2CH_3$

(m) 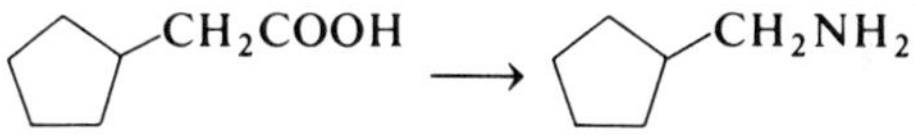 cyclopentyl-$CH_2COOH \longrightarrow$ cyclopentyl-CH_2NH_2

(n) 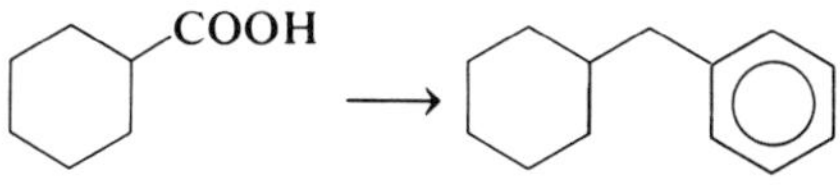 cyclohexane-COOH ⟶ benzylcyclohexane

(o) 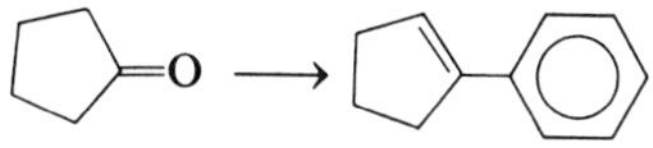 cyclopentanone ⟶ phenylcyclopentene

(p) 3-methylbenzaldehyde → 1-(3-methylphenyl)butan-1-ol

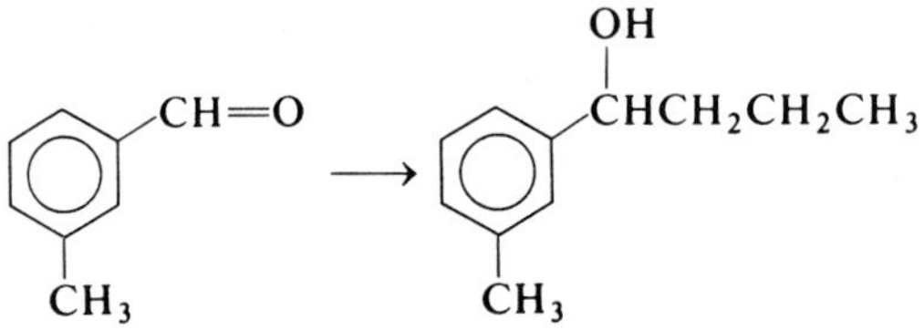

(q) Ph—CH(Br)CH₂CH₂CH₃ → Ph—CH₂—CH(OH)—CH₂CH₃

9. Indicate the more acidic compound of each pair and the most acidic hydrogen atom of that compound.

(a) Ph—COOH   CH₃COOCH₃

(b) 2-methylphenol   benzyl alcohol (PhCH₂OH)

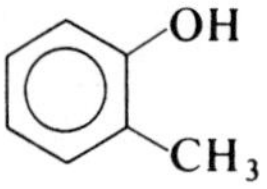

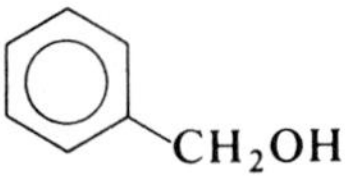

(c) 2-methylpyridine   toluene (Ph—CH₃)

(d) CH₃—C(=O)—CH₂—C(=O)—OCH₃   CH₃—C(=O)—CH₂—CH₂—C(=O)—OCH₃

(e) O₂N—C₆H₄—CH₃   CH₃O—C₆H₄—CH₃

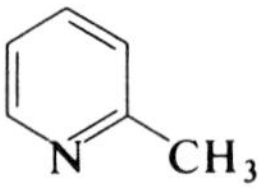

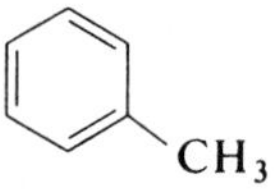

(f) bicyclic diketone   2-methyl-1,3-cyclohexanedione

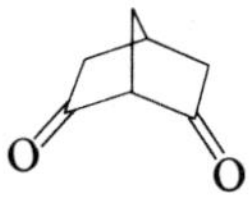

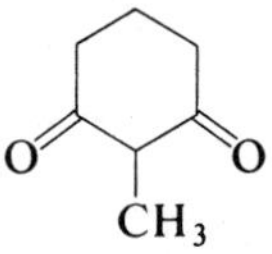

(g) CH₃COOCH₂CH₃   CH₃COOH

(h) Ph—CH₂—C(=O)—OCH₃   Ph—O—C(=O)—CH₂CH₃

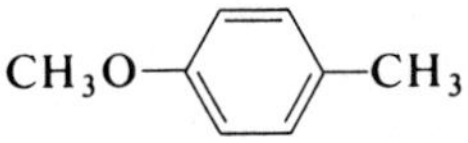

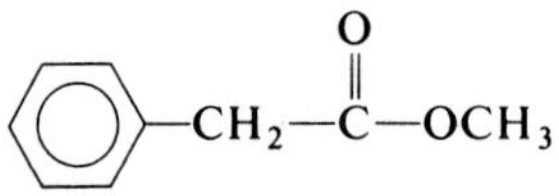

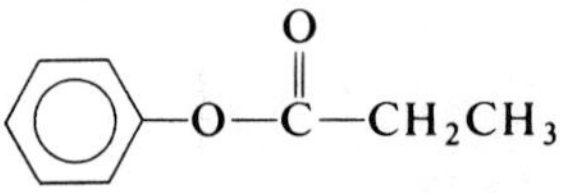

(i) ClCH₂COOH   BrCH₂COOH

(j) ⁻O—C(=O)—C₆H₄—NH₃⁺   CH₃O—C(=O)—C₆H₄—NH₃⁺

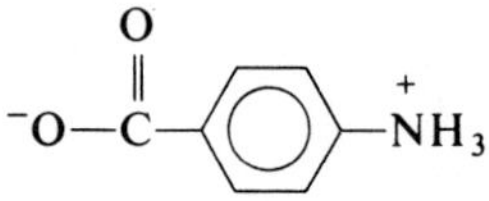

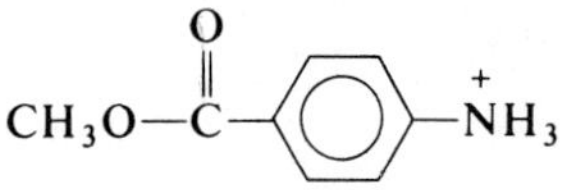

(k) 2-hydroxy-1,4-naphthoquinone   2-naphthol

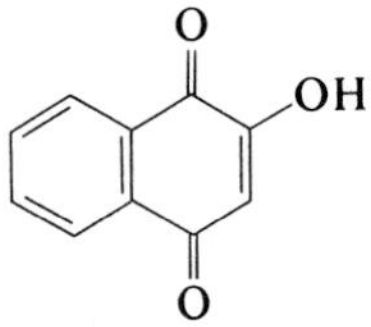

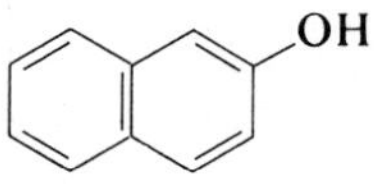

(l) ClCH$_2$CH$_2$CH$_2$COOH   CH$_3$CH$_2$CHCOOH
                                              |
                                              Cl

(m) O$_2$N—C$_6$H$_4$—COOH   CH$_3$O—C$_6$H$_4$—COOH

(n) CH$_3$—C$_6$H$_4$—$\overset{+}{\text{N}}$H$_3$ C$_6$H$_5$—$\overset{+}{\text{N}}$H$_3$

(o) CH$_3$NO$_2$ CH$_2$(NO$_2$)$_2$

(p) 3-Cl-C$_6$H$_4$-COOH   4-Cl-C$_6$H$_4$-COOH

(q) 6-O$_2$N-naphthalen-2-ol   1-O$_2$N-naphthalen-2-ol   (See Problem 10, Chapter 2, p. 97)

REFERENCES

1. (a) R. B. Woodward and R. Hoffmann, *Angew. Chem. Intern. Ed. Engl.*, **8**, 781 (1969); (b) R. Hoffmann and R. B. Woodward, *Science*, **167**, 825 (1970).

2. R. Stewart, *Oxidation Mechanisms: Applications to Organic Chemistry*, New York: W. A. Benjamin, 1964.

3. R. T. Morrison and R. N. Boyd, *Organic Chemistry*, 2nd Ed. Boston: Allyn and Bacon, 1966.

4. C. R. Noller, *Chemistry of Organic Compounds*, 3rd ed. Philadelphia: Saunders, 1965.

5. H. D. Weiss, *Guide to Organic Reactions*. Minneapolis: Burgess, 1969.

6. J. March, *Advanced Organic Chemistry: Reactions, Mechanisms, and Structure*. New York: McGraw-Hill, 1968.

7. L. F. Fieser and M. Fieser, *Advanced Organic Chemistry*. New York: Reinhold, 1966.

Index

Absolute configuration, 46, 47
Absorbance (A) definition, 71
Acenaphthylene, 18
Acetaldehyde, 80
Acetals, 22, 29, 123, 137
Acetamide, 63
Acetic acid, 33, 127, 132
Acetone, 70, 131
Acetylene, 13, 132, 133
Acetylenes, see Alkynes
Acid:
  anhydrides, 26, 31, 118, 120
  bromides, 25, 30
  catalysis, 102, 119. 121, 122
  chlorides, 25, 30, 62, 118, 120, 136–143
  salts, 30, 115, 120, 125–128, 130, 131
Acids, 91, 142
Acrylic acid, 33
Acyloins, 26, 32
Adamantane, 17, 58
Addition reactions, 102–108, 113, 122–125, 136–139
Adipic acid, 33
Alcohols, 22, 28, 41, 56, 63, 102–104, 109, 110, 112–114, 118–123, 131–143
Aldehydes, 25, 28, 62, 80, 105, 113, 114, 122–125, 132, 133, 135–142
Alkanes, 4, 9–13, 102, 107, 113, 114, 131–134, 140
Alkenes, 6, 13–14, 102–111, 131–134, 136–139
Alkynes, 6, 13, 14, 102–111, 132, 133, 136, 138
Allenes, 13, 111
Allyl group, 13
Amides, 25, 30, 56, 63, 117, 119–121, 132, 134–139
Amine oxides, 23, 32, 110. 137
Amines, 23, 29, 56, 63, 110, 112, 115, 120, 121, 123, 133–135, 137–139, 141–143
Amyl group, 12
Androstane, 43
Aniline, 81, 112, 115, 129
Anions, 39, 40, 108
Anthracene, 18, 61
anti-, 45–47
Antibonding molecular orbitals, 75, 76, 78, 80

Anti-Markovnikov addition, 104
Aromatic compounds, 7
  chemistry of, 111–114
Aufbau principle, 74
Auxochromes, 81
Azides, 23, 29, 41
Azo compounds, 23, 29, 80
Azoxy compounds, 23
Azulene, 18, 81

Balanced equations, 130–131, 140
Base catalysis, 119, 120, 122
Bases, 91
Benzene, 7, 15, 65, 76, 77, 81, 91, 111, 112
Benzoic acid, 33, 63, 66, 128, 132
Benzyl group, 15
Bicyclic hydrocarbons, 15–19
Biphenyl, 20
Boiling points, 2, 53–58, 95
Bonding molecular orbitals, 75, 76, 78, 80
Borane, 102, 104, 138, 139
Bornane, 22
Borneol, 43
Brackets, 15, 55
Bromine, 101. 102, 133, 136, 138
Brønsted Acid-base Theory, 125
1,3-Butadiene, 78, 79
Butane, 5, 10, 59, 60
Butanol, 56
sec-Butyl group, 12
tert-Butyl group, 12
Butyric acid, 33, 127, 135
γ-Butyrolactone, 38, 121

Camphane, 22, 58
ε-Caprolactam, 38
Carbanions, 39, 40
Carbenes, 37, 38, 106
Carbodiimides, 23, 32
Carbohydrates, 21
Carbon dioxide, 113, 114, 130, 132, 133
Carbonium ions, 37–40, 103, 110, 114
Carbonyl group, 25, 48, 55, 61, 62, 80, 83, 105, 113, 114, 118–120, 122–126
Carboxylic acids, 1, 2, 25, 26, 28, 33, 56, 63, 80, 105, 107, 110, 113–115, 117,

Index

118–121, 125–128, 130–133, 135–143
  dimers of, 1
Catalytic hydrogenation, 107, 134, 136, 140
Cations, 37–40
Cerium(IV), 129, 130, 135, 139
Chain reaction, 115–117
Characteristic groups:
  definition, 27
  list of, 28–32
Chemical Abstracts nomenclature, definition, 8
Chemical properties (Chaper 3), 101–143
Chemical shift, 91–95, 97–99
Chiral molecules, 46, 47
Chlorine, 115–117
Chromic acid, 135, 139
Chromophore, 80, 81
Cinnamic acid, 107
Cinnamyl group, 15
cis-trans Isomers, 4, 44–47, 57, 63, 81, 96, 107, 108
Clemmensen reduction, 134, 137
Color, 71–73, 101, 131
Column, Gas-liquid partition chromatography, 67–70
Conjugate acid, 125
Conjugate base, 125
Conjugated carbonyl groups, 80, 84
Conjugated double bonds, 73, 78, 79
Conjugated olefins, 106
Constitutional isomers, 43, 47, 48
Constructive interference, 76
Coupling constant, 91
Crotonic acid, 33
Crystal packing, 58, 59, 61
Cumene, 15
Cyanide, 41, 112, 121
Cyclic hydrocarbons, 14, 15
Cycloaddition reactions, 107
Cycloalkanes, 60
Cyclohexanone, 61, 83, 88, 89, 134, 135
Cyclohexene, 65, 101, 110, 134, 135
Cyclopropane rings, 7, 48, 106

Decalin, 19, 70
Decane, 10, 58, 59
Dehalogenations, 108, 109, 139
Dehydrations, 108, 110
Dehydrogenations, 110, 111
Dehydrohalogenations, 108, 109, 139
Delta (δ), 87
Destructive interference, 76
Detector, gas-liquid partition chromatography, 67–70
Deuterium oxide, 114
Diazo compounds, 23, 30, 80, 106
Diazonium ions, 37, 39, 112, 114, 138

Diborane, *see* Borane
Dichlorocarbene, 38
1,2-Dichloroethene, 57, 63
Dichloromethylene, *see* Dichlorocarbene
Diels-Alder reaction, 106, 107, 137
Dienophile, 106, 107
Diethyl ether, 55
β-Diketones, 126
2,4-Dinitrophenylhydrazine, 123
Diphenylmethane, 20
Diphenylpolyenes, 72
Dipole moments, *see* Dipoles
Dipoles, 55–58, 61–63, 65, 82, 84, 88, 113, 127
Disulfides, 24, 28
Divalent radicals, nomenclature for, 13
Docosane, 10, 59
Dodecane, 10
Dotted lines, 3
Double bonds, 6
Durene, 58

E-, 46, 47
Eicosane, 10, 59
Eigenfunction, 73
Electromagnetic radiation, 70–95
Electron density, 86–88, 90, 91, 93
Electron spin resonance, 53
Electronic configurations, 74–80, 97
Electronic spectra, *see* Ultraviolet-visible spectroscopy
Electronic transitions, 70–82
Electrophiles, 103, 108
Elimination reactions, 108–111
endo-, 45–47
Entgegen, 46
Epoxidation, 104, 105
Epoxides, 22, 28, 104, 105
Epsilon (ϵ), *see* Molar extinction coefficient
Esters, 25, 30, 62, 63, 105, 110, 115, 117–121, 134, 136–139
Ethane, 4, 10, 59, 87, 115, 116, 131–135
Ethene, *see* Ethylene
Ethers, 22, 28, 41, 62, 63, 115, 128, 132
Ethyl diazoacetate, 106
Ethylene, 13, 74-76, 78, 79, 131–133
Excited states, 70, 74, 115
Exit port, gas-liquid partition chromatography, 67
exo-, 45–47

Fats, 21
Field effect, 88
Fingerprint, 83, 85
Fluorene, 18

Index

Forbidden transition, 79
Force constant of a bond, 83
Formaldehyde, 113
Formic acid, 33
Formyl-group, 28
Free radicals, 24, 37–40, 115–117, 140
Friedel-Crafts reaction, 112, 138, 139
Fumaric acid, 33
Functional class name, definition, 27

Gas chromatography, *see* Gas-liquid partition chromatography
Gas-liquid partition chromatography, 67–70, 96
Geneva Rules, 8, 9, 33
Geometric isomers, *see cis-trans* Isomers
Glutaric acid, 33
Glycols, 23, 102, 132–134, 137
Greek letters, 15, 37
Grignard reaction, 112–114, 124, 137–139
Guanidine, 34

Halides:
  acyl, 25, 30, 41, 62, 112, 118, 120, 136–143
  alkyl, 26, 29, 41, 102, 103, 105, 108, 109, 111–117, 121, 124, 127, 133, 136–143
Halogenations, 116, 117, 136
Halonium ions, 37, 39
Hectane, 10
Helium, 67, 69
Hemiacetals, 122
Heneicosane, 10, 59
Heptane, 10, 55, 59
Hexacene, 18
Hexane, 10, 54, 59, 65, 66
Hexanol, 66
Hofmann degradation, 135, 138
homo-, 40, 42
Homotropylium ion, 42
Hund's Rule, 74
Hybrid orbitals, 4
Hydrazine, 134
Hydrazines, 23, 29, 123, 134
Hydrazones, 23, 123
Hydrogen, 131, 132, 134
Hydrogen bonds, 1, 56–58, 64–66
Hydrogen peroxide, 102
Hydroperoxides, 23, 41
Hydroxylamines, 23, 31, 110, 123, 133

Imides, 26, 31
Imines, 23, 29
Indene, 18
Induced magnetic fields, 86, 92

Inductive effect, 86, 90, 127
Infrared frequencies for common bonds, 84
Infrared spectroscopy, 70, 81–85, 97–99
Initiation steps, 116, 117
Injector block, gas-liquid partition chromatography, 67
Injector port, gas-liquid partition chromatography, 67
International Union of Chemistry (IUC), 8
International Union of Pure and Applied Chemistry (IUPAC), 8
Iodoform, 136
Iodoform reaction, 135, 137
Iodoso compounds, 26, 29
Iso-, 12
Isoamyl group, 12
Isobutane, 5, 12
Isobutyl group, 12
Isocyanates, 27, 29
Isocyanides, 27
Isohexane, 12
Isohexyl group, 12
Isomers, definition, 5, 6
Isopentane, 12
Isopentyl group, 12
Isophthalic acid, 33
Isopropenyl group, 13
Isopropyl group, 12
IUPAC, *see* International Union of Pure and Applied Chemistry
IUC, *see* International Union of Chemistry

J, *see* Coupling constant

Kekulé resonance structures, 7
Ketals, 32
Ketenes, 27, 34
Ketones, 25, 28, 41, 62, 105, 113, 114, 122–125, 132, 134–141
Ketyls, 39, 40

Lactams, 37, 38, 121
Lactones, 37, 38, 121, 135
Lambda max (λ_{max}), definition, 71
Lauric acid, 33
Lead tetraacetate, 135
Leaving groups, 108–110, 115, 118, 119
Liege Rules, 8
Linear Combination of Atomic Orbitals (LCAO), 74, 75
Lithium aluminum hydride, 134, 138, 139
Lithium reagents, 124

Index

Magnetic anisotropy, 91–93
Magnetic field, 85, 91, 92
Maleic acid, 33
Malonic acid, 33
Markovnikov's Rule, 103, 108
Mass spectroscopy, 53, 70
Melting points, 2, 58–64, 81, 95–97, 124
p-Menthane, 22
Menthol, 43
Mercaptans, 24, 28, 63, 115
Mesitylene, 15
meta- (m-), 15
Methacrylic acid, 33
Methane, 10, 59, 116, 132, 133
Methanol, 66, 115, 132, 133
Microwave radiation, 71
Molar extinction coefficient (ϵ), definition, 71
Molecular flexibility, 2, 54, 55, 58, 64
Molecular shape, 2, 4, 5, 54, 55, 58, 64
Molecular size, 2, 54, 55, 58, 64
Molecular weight, 2, 54, 55, 58, 64
Molozonide, 105
Multiple bonds, 6
Myristic acid, 33

Naphthacene, 18
Naphthalene, 18, 81, 131
Naphthalene radical anion, 131
Naphthoic acid, 33
Neo-, 12
Neopentane, 12, 54, 55
Neopentyl group, 12
Nitrates, 23
Nitriles, 23, 30, 41, 80, 121, 132, 134
Nitrites, 23, 112
Nitro compounds, 23, 30, 62, 80, 112, 128, 129, 133, 134, 142, 143
Nitroethane, 56
Nitroso compounds, 23, 30, 133
Nitroxyl radicals, 24
Nodes, 76, 78, 79
Nomenclature:
   absolute configuration of asymmetric carbon atoms, 47
   additive, 40, 42
   bridged hydrocarbons, 15–19
   cis-trans isomers, 43–47
   conjunctive, 42, 44
   cyclic hydrocarbons, 14, 15
   geometric isomers, *see cis-trans* isomers
   hydrocarbon ring assemblies, 19, 20
   hydrocarbons, 8–21
   ions, 37–40
   lactams and lactones, 37, 38
   organic compounds that possess functional groups, 27–43
   organic molecules (Chapter 1), 1–54
   phosphorus compounds, 42, 44
   polycyclic systems, 17
   radicals, 37, 38, 40
   radicofunctional, 40, 41
   replacement, 42, 44
   saturated acyclic hydrocarbons, 9–13
   spiro hydrocarbons, 19, 20
   substitutive, 28–37
   subtractive, 40, 41, 43
   terpenes, 21, 22
   unsaturated acyclic hydrocarbons, 13, 14
Nonane, 10, 59
Nonbonded electrons, 115
Nonbonding molecular orbitals, 80
Nor-, 40, 41, 43
19- and A-Norandrostanes, 43
Norbornane, 16, 22
Norbornanol, 43, 45
Norbornene, 22
Norborneol, *see* Norbornanol
Norbornyl cation, 39
Norcarane, 22
normal- (n-), 12
Nuclear magnetic resonance, 70, 85–95, 97–99
Nucleophile, 114, 115
Nucleophilic substitution reactions, 114
Numbering rules, 10, 11

Octane, 10, 54, 57, 59
Olefins, *see* Alkenes
Optical density, definition, 71
Orbitals, 4, 73–80
ortho- (o-), 15
Osmium tetraoxide, 135, 136
Oxalic acid, 33
Oxidation, 129–140
Oximes, 23, 31, 63, 123
Oxonium ions, 37, 39
Ozone, 105, 140
Ozonides, 105
Ozonolysis, 104, 105, 136

Palmitic acid, 33
para- (p-), 15
Paraffins, *see* Alkanes
Pauli's Principle, 74
Pentacene, 18
Pentacontane, 10
Pentadecane, 10
Pentane, 10, 54, 55, 59
tert-Pentyl group, 12
Peracids, 26, 32, 105, 135–137
Peresters, 26
Permanganate, 135, 140

Peroxides, 23, 26, 30, 41
Peroxycarboxylic acids, see Peracids
Phenanthrene, 18, 61, 111
Phenethyl group, 15
Phenols, 126
ortho-Phenylene, 15
Phenyl group, 15
Phenylhydrazine, 123
Phosphates, 26, 44
Phosphine oxides, 26, 44, 124
Phosphines, 26, 44, 124
Phosphites, 26, 44
Photochemistry, 106, 107
Phthalic acid, 33
Physical properties (Chapter 2), 53–100
Physical property, definition, 53
Pinane, 22
Planck's constant, 71
Poisoned catalyst, 107
Polynuclear aromatic hydrocarbons, 18
Principal group, definition, 27
Probability, 73
Propagation steps, 116, 117
Propane, 4, 10, 59, 60, 62, 116
Propionic acid, 33
Proteins, 21
Pyridine, 88–90
Pyrolysis reactions, 110

Quaternary ammonium ions, 23, 110
Quinone, 34, 132

R-, 47
Radical anions, 39, 40, 131
Radical cations, 131
Radical ions, 37–40
Radicals, see Free radicals
 alkyl, 2, 10
Radicofunctional nomenclature, definition, 40
Recorder, 67, 69
Recrystallization, 124
Rectus, 47
Redox reactions, see Oxidation and reduction
Reduction, 129–140
Resonance, 1, 4, 7, 55, 62, 89, 90, 97, 113, 125, 126, 128, 129
 effect, 88, 90, 127, 128
Restricted rotation, 44
Retention times, 69, 96
Ring assemblies, definition, 5
Rotational transitions, 71, 76, 77
Rubicene, 19

S-, 47
Saturated compounds, 6, 9–13

secondary- (sec-), 12
Selection rules, 79
Selenium dioxide, 135
Semicarbazide, 124
Semiquinones, 39, 40
Sequence rules, 46, 47
Shielding, 86
Simmons-Smith reaction, 106, 136
Sinister, 47
Solubility of organic compounds, 64–67, 96
Spin, see Nuclear magnetic resonance
Spiro:
 atom, definition, 19
 compounds, 5
 hydrocarbons, 19, 20
Stationary liquid phase, 67
Stationary solid phase, 67
Stearic acid, 33
Stereoisomers, 43–48, 50, 51
Steroids, 22, 41
cis and trans-Stilbenes, 81
Strain energy, 7, 22
Styrene, 15
Substituent, definition, 27
Substitution reactions, 111–121, 136–140
Succinic acid, 33
Sugar, 65, 66
Sulfides, 24, 28, 41, 62, 63, 115
Sulfinic acids, 24, 29
Sulfonamides, 24, 29
Sulfonate esters, 115
Sulfones, 24, 28, 41, 62
Sulfonic acids, 24, 29, 115
Sulfonium ions, 37, 39
Sulfoxides, 24, 28, 41, 62
Sultams, 37, 38
Sultones, 37, 38
Symmetry, 60, 63
syn-, 45–47
Synthesis, 101, 141, 142
Systematic nomenclature, see Geneva Rules

Terephthalic acid, 33
Termination steps, 116, 117
Terpenes, 21, 22
tertiary- (tert-), 12
Tetracosane, 59
Tetradecane, 10
Tetralin, 19
Tetramethylsilane, 55, 87
Tetrapentacontahectane, 10
Thermal conductivities, 69, 70
Thermal conductivity detectors, gas-liquid partition chromatography, 69
Thermistor, 69
Thiols, see Mercaptans
Toluene, 15, 130
Translational energy, 70, 82

Triacontane, 10, 59
Tricosane, 10, 59
Tridecane, 10
Triple bonds, 6
Trityl group, 15, 38
Trivial names, 9, 10, 12, 13, 32, 33, 37, 40
Tropylium ion, 39

Ultraviolet-visible spectroscopy, 70–82, 96, 97, 102
Undecane, 10
"Unofficial" nomenclature, definition, 9
Unsaturated compounds, 6, 13, 14
α, β-unsaturated ketones, 84, 89
Unsaturation, 111
Ureas, 26, 29
Ureid-o, 29
Urethans, 26

Valeric acid, 33, 114
δ-Valerolactam, 121
Vapor phase chromatography, *see* Gas-liquid partition chromatography

Vapor pressure, 54, 67–69
Vibrational transitions, 71, 76, 77, 82–85
Vinyl group, 13

Wave equation, 73
Wave function, 73, 75, 76, 79
Wave mechanics, 3, 73, 89
Wedges, 3
Wittig reaction, 124, 137
Wolff-Kishner reaction, 134, 138
Woodward-Hoffmann selection rules, 107

X-ray crystallography, 53, 59
Xylene, 15

Ylides, 124, 137

Z-, 46, 47
Zinc amalgam, 134, 137
Zinc-copper couple, 106
Zusammen, 46

QD
253
.T67
1971

72-3185

Trahanovsky, Walter S
Functional groups in organic
compounds

DISCARDED

NOV 20 2024

Asheville-Buncombe Technical Institute
LIBRARY
340 Victoria Road
Asheville, North Carolina 28801